RAMOND BURGIN
BIG STONE GAP, VA., AUG. 1990
(FROM JOHNNY)

COME ALL YE BOLD MINERS

Come All Ye Bold Miners

Ballads and Songs of the Coalfields

Compiled by

A. L. LLOYD

LAWRENCE AND WISHART LTD
LONDON

Lawrence and Wishart Ltd
39 Museum Street
London WC1A 1LQ

First published by Lawrence and Wishart, 1952
This new, revised and enlarged edition published in 1978

Printed and bound in Great Britain at
The Camelot Press, Ltd, Southampton

CONTENTS

PREFACE

The first edition of *Come All Ye Bold Miners* appeared long ago, in the dear dead days of the Festival of Britain. In the spring of 1951, as part of the mining industry's contribution to the Festival, it was decided to make a collection of coalfield songs before they disappeared. By way of money and encouragement, little enough was forthcoming. The masters of the industry believed the enterprise would be in vain. If there had ever been such songs, they were no longer recoverable. Should an attempt be made, it would have to be on the cheap. I was asked to arrange a competition, and by means of a cinema newsreel of mining affairs and the National Coal Board's magazine, *Coal*, miners were invited to submit any songs they knew of the life, work, pastimes, disasters and union struggles in the coalfields. Modest prizes were offered for those songs judged to be the best 'finds'.

There were more than a hundred entries. Not all the songs submitted were what we understand as folk songs. Some were parodies, recitation pieces, parlour ballads of the type of *Don't go down in the mine, Daddy*. Many were stage songs of the past, particularly from the Tyneside music halls of a century and more ago. But a number of fine songs not hitherto seen in print came to light. These, and a few songs taken directly from the singing of pitmen, formed perhaps the most valuable part of the first edition, which otherwise was compiled from printed sources – garlands, ballad sheets, leaflets – and from miners' manuscripts.

As it turned out, that modest first edition, skimpy in content, poorly edited as it was, had an influence far beyond expectation. To some pitmen, especially young ones enthusiastic for the folk song revival, *Come All Ye Bold Miners* was 'the Bible'. The compilation had the effect of restoring to vigorous life many past songs, stimulating investigators to seek out lyrics dormant in cold corners of the memory of old miners or gathering dust in library cupboards, and best of all, encouraging members of colliers' families to chance their arm at making songs for themselves about their own lives. This second edition reflects that upshot; it is much bulkier than the first edition because so many more old songs have come to light, and so many more new ones have been made up in the intervening quarter-century or so, also we all know much more about the subject nowadays. Industrial folklore, thought to be a will o' the wisp, is taking on the proportions of an academic discipline as well as a cultural force.

In the first edition, I gratefully acknowledged the contributions of the many mineworkers who gave their songs to me, or who sent me fruits of their researches, notably J. S. Bell, a coal-face driller of Prescot, Lancs, whose investigations of the balladry of the 1844 Durham strike have been so fertile. A bow, too, to the shade of W. H. Armstrong of Tanfield Lea, Co. Durham, for generously allowing me to use a number of ballads made by his father, the now celebrated Tommy Armstrong, who was wellnigh unknown in 1952 when the first edition was published. Besides these, for this new edition I signal my gratitude to the Elliott family of Birtley, Co. Durham, splendid bearers of the traditional folklore of mining, to 'Jock' Purdon, formerly deputy at the 'Cotia Pit (famous in song if not particularly lustrous in the history of the coal trade), likewise to Bert Draycott, miner, of Stockton-on-Tees, the Tommy Armstrong of our time. George McIntyre of Glasgow gave me good help with Scottish stuff, and Mike Yates was most generous in passing on to me the rich broadside harvest he found in the Sheffield University Library. I owe a special debt to Frank Rutherford of Newton Hall, Co. Durham, who amiably drew my attention to many broadsides and manuscript texts, and in doing so gave an example of editorial precision that, I fear, I often fail to live up to.* Finally, to all the singers and songmakers and purveyors of coaldust lyrics mentioned in the Notes at the back of this volume, I offer my best thanks.

A. L. Lloyd
April 1978

* À propos, I have made no attempt to standardize the rendering of dialect. Particularly this is so with the songs from Co. Durham. I have my excuses. Some of the texts are quoted from print or manuscript where the dialects are already put in sundry forms that it would be unseemly to alter; some – notably those by Alexander Barrass and Tommy Armstrong – are set down by their authors in an imitated pronunciation whose contortions make for hard reading and I have been constrained to simplify them; in any case, as transcription from recording shows, dialect usages may vary within a single pit village, and indeed successive recordings of the same song from the same singer may show differing degrees of 'pitmatic' or other dialect. So: inevitable disorder.

INTRODUCTION TO THE SECOND EDITION

These are rough songs, mostly made by rough men. Hardly any have much claim to poetic art. Only a few are folk songs in the strictest sense – particularly in the sense of being so shaped and eroded in passing from mouth to mouth that they acquire the smoothness, subtlety and surprise of a sea-pebble. For one reason and another, tradition has hardly got to work on the pieces included here; they remain as coarse and awkward as the day they were first sung.

Yet they are songs that have meant much to coal miners. Their very hoarseness is a virtue in as much as the voice that speaks in them of work and joy and disaster and struggle is the voice of the miner himself or of someone sharing his life. The 'beautiful' is not the point here, rather the 'good'. A 'good' song in this context is one that positively affects the community for whom it is destined, that makes for a clear reflection of circumstances, that aids understanding of common plight. If these songs, acclimatized to coal and stone and the dust of the mainway, have not much literary felicity, they have yet proved, some of them, that they have the power to move mountains.

Songs that are so close to the heart of the common people may have a functional quality that goes far beyond mere diversion. Many of them were made up not so much to decorate life as to make it bearable. In an East European folk song a peasant woman warns her daugher: *Learn these songs and learn them well, for you never know what tomorrow may bring.* Such advice would be taken to heart by the oldtime collier's wife, struggling to make ends meet on pit pay, and never free from the threat of strikes and lockouts or the sudden terror of the calamity-whistle. Some of these songs, then, have not been sung for singing's sake but rather to lift a weight off the heart. Such songs as *The Collier Lass, The Coalowner and the Pitman's Wife, Brave Collier Lads, The Durham Lock-out, The Banks of the Dee, Farewell to 'Cotia* may show the mettle of a community clearer than a shelf of history books.

The origins of the pieces in this collection are various, but there is a good proportion of oral stuff that tends to be more vivid and less windy or cliché-ridden than the printed popular literature. It is as industrial folklore that these songs are unique. It is doubtful if any other industry in Britain has such a body of lyric related to the job itself, or to the life, diversions and struggles of the people engaged in that job. The coal miners have always tended to be men apart. The fact that they worked underground away from the light of sun, and their ragged soot-faced appearance as they filed along the roadsides at the day's end, caused them to be looked on as a strange race, dirty, savage, prodigal and drunken, not worthy of education nor of any better conditions than the mine-owners chose to grant them.

The coal miners of Britain were a creation of Elizabethan times, when the air was full of enterprise and capitalism was young and bold. Till then, men who dug coal did so only as a sideline to their regular work as farm labourers and smallholders. On the Tyne and Wear and elsewhere the Church had owned most of the mines, but after the Reformation the pits were taken out of the bishops' hands and leased by the Crown to merchant adventurers, with the result that the coal trade expanded greatly. Within two generations between 1565 and 1625 coal exports from the Tyne rose from 35,000 tons to 400,000. By the end of the seventeenth century the Gateshead and Whickham pits alone were producing 100,000 tons a year. The number of miners rose as the industry expanded, and with the transformation effected by the Industrial Revolution the collier emerged very much as we know him today – wary, sharp-witted and blunt of tongue, strong set against rebuff and little inclined to lick the boots of any authority.

Conditions, always hard in the coalfields, grew harder as the pits were driven deeper. Along the Tyne, the Wallsend Main was won at 100 fathoms in 1781, Hebburn followed at 129 fathoms in 1794, Percy Main (120 fathoms) in 1802, Jarrow (128 fathoms) in 1803, South Shields (140 fathoms) in 1810. In Scotland the pitmen were bound to their master in quasi-slavery for life, being by law 'the property of their landlords, appurtenances to their estate, and transportable with them to any purchaser'. With the industry expanding so fast, the noble owners wanted to prevent their colliers from moving into new pits that might offer higher wages, and their power to retain their workers as serfs was confirmed and extended by statutes which were not abolished until the end of the eighteenth century. And, says J. U. Nef, the historian of the rise of Britain's coal industry, 'while slavery did not develop in the chief coal-producing districts outside Scotland . . . the miners there do not appear to have enjoyed a much higher social position or a much greater measure of independence than their Scottish fellows'. A foreign visitor to the Tyne was struck with the number of notices inserted in local newspapers by mine-owners offering rewards for knowledge of the whereabouts of runaway miners, and threatening to prosecute whoever might employ them.

To the masters, the colliers of the time seemed 'a most tumultous set of sturdy people, greatly impatient of control, very insolent and much void of common industry'. A correspondent wrote to William Pitt: 'When I look round and see the country covered with thousands of pitmen, keelmen, waggonmen and other labouring men, hardy fellows strongly impressed with the new doctrine of equality, and at present composed of such combustible matter that the least spark will set them in a blaze, I cannot help thinking the supineness of the Magistrates very reprehensible.' To the Reverend Thomas Gisborne, writing of colliers near Manchester in 1798, the men seemed 'drunken, profane, and full of supposed grievances'.

Not all coalowners were so uncharitable. The benevolent Mrs Montagu of Tyneside found the pitmen working in her colliery were strange, even savage, but not entirely unlovable. She remarks: 'As to Denton, it has mightily the air of an ant-hill; a vast many black animals forever busy. Near fourscore families are employed on my concerns here. Boys work in the colliery from seven years of age. I used to give my colliery people a feast when I came hither, but as the good souls (men and women) are very apt to get drunk, and when drunk very joyful, and sing and dance and holloo and whoop, I dare not . . . trust their discretion to behave with proper gravity; so I content myself with killing a fat beast once a week, and sending to each family, once, a piece of meat. It will take time to get round to all my black friends.' So might an indulgent slave-owner have written from a pillared mansion in Virginia.

But in a letter to a London paper a writer of Mrs Montagu's time gives another picture. 'To be sure it is the business of the proprietors to represent them as a set of lazy, disorderly fellows who want only to increase their wages for the sake of extending their extravagances; the more they are kept down, the more their masters will be enabled to venture ten thousand guineas on a favourite horse or the accidental turn of a card. But the sensible part of the kingdom must see that when butter in the northern part of England is at sixpence, and butcher's meat at threepence a pound, the man who has but seven shillings a week to support himself, a wife and four or five children, can have no mighty matter to squander away at an alehouse.'

Such is the background to the older songs in this volume. With the coming of steam, industrialism spread so that by the early years of the nineteenth century much of the once-green landscape was now bare and black, with a tough thin-herb folk growing on it among the coal-tips. In many parts the new Iron Age meant the ruin of traditional ways and the disappearance of much that had given fire and zest to the working man's life. The former farm-hands, now operatives in some long, dark factory. were bewildered and downcast, without much heart to sing. But in the coalfields the pitmen seemed to exult in the new age. The blackness of it was their element, a natural part of the landscape. For them, the stir of industrial life, with its stresses and struggles, its new techniques and fresh horizons, its battles against nature below-ground and against the militia at the pithead meant a change but not a break in tradition; they accepted it and gave voice to it out of their coal-stained mouths. The ranting colliers who had sung of meeting the devil in the pit and knocking off his horns with a coal-pick made new songs about hard times and flinty owners and the need for combining and standing fast against the rigours of the age.

The temper of the miners may be read by signs: scars in the subterranean rock, myths and proverbs and anecdotes without end, records of the courts of justice, union minute-books, and songs still remembered. Walking down the dusty ways towards the face, the youngsters might send their voices echoing to:

> Byker Hill and Walker Shore,
> Collier lads for ever more!

A lamp directed on a beam might show a derisive stanza chalked by some wag to mock non-union miners:

> Join the union while you may.
> Don't wait till your dying day,
> For that may not be far away,
> You dirty blackleg miners.

The young girls working on the screens might grin across at the boys and cry:

> My mother said that I could be a lady
> If from my collier lad I'd quickly part.
> I'd rather walk the bottom of the ocean
> Than I'd give up my collier sweetheart.

And a crouched hewer cutting into the black face might grunt through his teeth, with a pick-blow falling at the end of each line:

> She's a big lass an' a bonny lass,
> An' she likes hor beor,
> An' they caal her Cushie Butterfield
> An' Aa wish she wes heor.

Human history is work history; and many of the songs that working men have evolved for their own use are valuable documents for understanding what has made men move in the past and what will make them move again.

Coalminers' songs mean something special in the realm of industrial lyric because the pitman's job is on the threshold between basic production and industry; after all, well on in the twentieth century nearly all coal was still hand-hewn. The country workers who were the chief bearers of the old folk song tradition had lived in a world where a man's calling had a wellnigh sacred sense, and there was little tension between work and life, job and family. Something of the near-sacred had obtained among miners, but it was with them that the sharp separation of home, family and workplace first occurred which now has become normal in industrial society. The separation is deepened by shift-work, that destroys the natural rhythm of day and night, a rhythm already blurred by the darkness of the mine, and submits man to a measure of time artificially designed for productivity.

Well, as it says in the *I Ching*, 'Times change and with them their demands'. New social groups emerged amid radical upheavals as the working class began to take shape out of the rubble of the old 'peasant' and artisan classes, and inevitably new viewpoints were taken up. So into the domain of folklore comes a fresh factor, the labour movement, which found its own expression in songs, tales, customs and speechways. It has to be said that the folklorists, deep in their rustic trance, have been slow to approach the problems of the proletariat and to concern themselves with matters of industry, urbanization, mechanization and their effect on the homemade culture of people not only in the traditional trades but also in new realms of production.

We say 'folklore', folk song', but nowadays the terms are vague and the frontiers between tradition and innovation, communal and personal expression, are ill-defined. Anyone wishing to study workers' song in Britain has to deal with at least four kinds of composition. Associated with the proletariat are revolutionary anthems such as the Chartist

Song of the Lower Classes, Jim Connell's *Red Flag*, or the *International*, pieces representing a generalized, more or less global view, such as you would find in the song-books of the Socialist movement; the texts of such songs have an agitational content and an emotionally elated style that is connected with political writing over the last 150 years, but is sharply different from the style of folk song. Likewise there is a class of songs, now seldom sung, made by worker poets according to literary, often Shelleyish models, as like as not apostrophizing the proletariat as 'ye sons of toil', parodied by the workers themselves as 'ye tons of soil'. Then there are songs made on behalf of the working class by professional 'outsiders', and where Britain is concerned these carried more weight among the proletariat in the days when the workers' choral movement was strong (since the 1950s, formal choral performances have been much eclipsed by the informal, mainly solistic presentations made fashionable in the 'folk song revival' clubs). None of these kinds of songs has much of the character of folklore, and in the main we are little concerned with them in this collection. Our care is rather with songs made originally within, and for circulation among, the in-group of workmates and their families, more or less agreeing in style with traditional folk song in creation, shape and destiny, and generally able to flash a valid enough passport to be permitted across the vague frontiers of folklore, provided the controllers are not stiff-necked. Such songs are of two main sorts. The first layer accumulated during the early stages of working class formation, when the new workers' folk songs were often attached to the former country tradition and, directly or not, took over its musical and poetic means. These songs arose spontaneously in working-class communities and in the main were transmitted orally, and indeed this kind of composition has rather taken on a new lease of life, paradoxically it might be thought, since the end of the Second World War. The other layer of workers' folk songs shows pieces in which individual creation is more noticeable, and borrowings from the world of book poetry and conventional music are fairly clear. These pieces are not only orally spread but have become current to a marked degree through broadsides, workers' newspapers and periodicals, song-books, and cyclo-styled or handwritten copies.

It is natural that the character of workers' songs should be closely joined to working class history and should reflect sundry phases of the developing consciousness of the proletariat. So the early songs, often associated with dance, tend to happy-go-luckiness, even if luck was as scarce as happiness in the beginning period of massive industrial conversions. Unlike their brothers on the Continent, British miners do not seem to have had a great deal in the way of pious song, and the fact is all the more surprising when we reflect on the close connection between Primitive Methodism and early trade unionism. Most of the nineteenth-century miners' leaders in Northumbria and the Midlands belonged to the sect (in Wales, it was the Baptists who were favoured) and learnt not only their oratory but also their organizational skill through the chapel. Towards the middle of the nineteenth century we find a certain staunch idealism in the songs of combination, that increases as the unions began to attract mass membership. In the latter part of the century the songs offered more and more realistic, often sardonically humorous pictures of the miner's life, underlain with sharp protest though seldom of conscious revolutionary tendency. More recent creations roam a broad and varied landscape, raising voices from below ground (*Putter's Lament*) to the surface (*Man Called Len*), from pub (*Shovellin' Back the Slate*) to union meeting (*Pound a Week Rise*), and offering a last salute to collieries being dismantled (*Farewell to 'Cotia, Lament for Albert*).

Dance-songs, squibs, parodies, non-narrative lyrics make up a sizeable part of the workers' song repertory, but the appearance of the full-fledged workers' *ballad* came quite suddenly, from the 1830s on. Our rural folk ballads usually reflected the world of smallholders, lone producers, free enterprisers. Personal emotion counted for much. Our industrial ballads show another world, with other preoccupations, where cares are displayed in terms of the community rather than the individual. If in the early stages of the industrial song tradition we have love ballads lying close to rural folklore, such as *The Collier Laddie, Jolly Joe, the Collier's Son*, and *Brave Collier Lads*, more characteristic compositions soon appeared, and we find ballad-like descriptions of work at the pits covering a wide span of time, as in *The Collier Lass, Five in the Morning, Oakey's Keeker, The Collier Lad is a Canny Lad*. More strikingly, in the course of time the pit-disaster ballads tended to pass from more or less personal tragedies to communal dramas and thence to themes of class struggle. From such pieces as *The Blantyre Explosion* and *Johnny Seddon*, through *The Colliery Explosion at Patricroft, Wigan*, the ballads of Starlaw, Trimdon Grange, Donibristle, Gresford, on to *The Watter's Brokken In* we find an increasing concentration on exposition and subsequent action, with a tendency to describe facts and finally to criticize them.

The injured and dead are seen not as toys of fate, as in the old folk ballads, but as victims of greed and neglect. Such ballads are not mere accusations of destiny and, more important, they cease to be tragedies of individuals. The concern is to tell the truth about events, the more so if it is falsified by the masters and their press. Such ballads do not work by revolutionary appeals like the artistic proletarian anthems such as the *International*; their aim is to arouse emotion against injustice, to strengthen the sense of solidarity, to move beyond lamentation over the workers' plight to burning protest against their oppressor. The realistic pattern of the serious mine ballad is so firmly fixed that the symbolic epic *Row between the Cages* has to be seen as an isolated masterpiece.

It will be noticed that most of the songs in this collection come from the English North-East. So far, the collecting of industrial folklore has been superficial and lopsided and by chance we know more about the Northumbrian miners than others. But is it by chance? Perhaps our north-eastern miners have been the most prolific creators and the best maintainers of this kind of songs. Certainly their territory has an unusually strong tradition of local and occupational lyric, reprinted time and again on broadsides, in garlands, songbooks and sheet-music albums, to say nothing of the repertory of oral songs that have not seen any light of print. During the nineteenth century Tyneside was rich in local ballad-makers, most of them working men. Not all were pitmen; for that matter not all were particularly sympathetic to the miners whom they pictured as lively but disreputable. Henry Robson, who wrote *The Colliers' Pay Week*, was a printer. J. P. Robson, author of the evergreen *Pawnshop's Bleezin'*, was a mechanic turned schoolmaster. Robert Nunn, one of the best of these humble authors, and one of those nearest to folk tradition (as in the *Sandgate Wife's Nurse Song*, for instance), was a slater who, blinded in a fall from a roof, contrived a living by fiddling and singing at miners' meetings and women's Benefit Club shareouts and in the public houses on pit pay nights. Other celebrated balladeers were former pitmen who, through hard times or as a result of accidents, were obliged to take to singing for a livelihood, graduating from the pub floor and mechanics' hall to the vaudeville stage. Such a man was George Ridley, a former miner in the Goose Pit, and later an employee at Hawks's Ironworks, where he met with an accident which unfitted him for heavy work. He became famous as a stage comedian and song-writer (*Blaydon Races*, the 'national anthem of the North-East', was one of his successes; *Cushie Butterfield* was another), but he never recovered from his accident, and after a public career of barely five years he died at the age of thirty in 1864.

If some of the songs in this collection are the work of humble professional writers, most of the others are made by working pitmen, known or unknown. Perhaps the most talented and genial of these miner-balladeers was Thomas Armstrong, the Tanfield poet. He was a small man with bowed legs. He was put to work at the age of nine, and his elder brother would carry him pickaback to the pit because of the pain of his crooked limbs. He worked all his life in the pits at Tanfield Moor, Tanfield Lea and East Tanfield, and was known through Durham as one always ready to turn his talent to raising sympathy or money for causes such as strikes, evictions, pit disasters. Though much of his work is lost, certain of his songs have passed into traditional currency and occur in various forms. W. H. Armstrong, the eldest of his sons, declared: 'There is no doubt my father was the greatest Genius of his day. He could and did stand up and without preparation compose songs. What would he have done had he had the education? Drink was his downfall. All his songs were based on facts. He was of all men the wittiest.' Another son remarked: 'His misfortune was to have fourteen children and a bottomless thirst. Me dad's Muse was a mug of beer.' And he recalls his father saying: 'When you're the Pitmen's Poet an' looked up to for it, wey, if a disaster or a strike goes by wi'oot a song from you, they say: What's wi' Tommy Armstrong? Has someone druv a spigot in him an' let oot aal the inspiration?'

Some of the best songs in this collection are the work of anonymous miners with no apparent facility or practice in the making of songs. It goes to the heart, the thought of the pitman stirred by the drama of some strike or disaster, who sits by candlelight with a blunt pen in his fist, staring at a piece of paper on which he has written the opening phrase: 'Come all ye bold miners . . .', and who wrestles by scratch and score with his stubborn muse till day dawns and the pit buzzer blows and another ballad has come bawling or timorous into the world.

In the previous (1952) edition of this book I had ended my Introduction with these words: 'Here, then, are the pit songs of the past. We await the pit songs of the future. If this humble collection can encourage one miner to make one ballad out of his working life, it will have achieved its aim.' Well, this new edition has nearly three times as many songs as the old one, and a good number of them have been made since the book first appeared, and it would not be vainglorious to suggest that some of these were sparked off by the appearance of that first volume of coalfield lyrics. The British establishment is good at absorbing and denaturizing those who aspire from below (the plight of a D. H. Lawrence is not so very different from that of the trade union leader raised to the peerage), and it is hard for a local or submerged culture to survive in the face of the entertainment corporations. Nonetheless, the song collection that is in your hands bears witness to a minority tradition that has not lost all its creative force, and indeed may yet surprise us all.

I

The Miner at Work

I

THE MINER AT WORK

What does the miner sing? He sings much the same songs as other working men: popular song-hits and parlour ballads, a few hymns, one or two standard opera airs and, if he is a Welsh miner, some bits of nineteenth-century oratorio. Besides that, he may have a few – very few as a rule – of what might be called work songs.

Mining is often a cramped job and the man cutting coal may have little breath to spare for singing. Where miners' work songs exist, they are not likely to be of the kind a man would swing a pick to. Rather they are songs *about* work, for singing out of the mine, in a public house perhaps, or in a pit-cottage parlour when friends have called, or at a union social, or on a coach-outing with workmates and wives to some green place out of sight of the coal-tips. Probably such songs are better called 'workers' songs'.

The songs the miner sings of his work are of many kinds. Most have been painfully made by the colliers themselves, though some have been written by authors who never wielded a pick, either wet or dry, in earnest. Sometimes the miner trying his hand will take a leaf out of the professional song-writer's book, and compose a piece full of high sentiment and moral reflection, but as a rule the genuine pit-songmaker prefers hard fact and plain statement, as in *Jowl and Listen* whose words mean much to the in-community of pitmen.

Of the songs in this section, the *Colliers' Rant* is the most deeply traditional, and *Down in a Coal-Mine* is closest to the stage. Both are well-known to miners, though nowadays it is the *Rant* which is learnt from print and kept for formal occasions, and the stage song which is passed on by word of mouth and brought out whenever spirits run high. Probably the Elliott family's spirited version of *Little Chance* has become the widest known of all these pieces outside the coalfields, but among miners themselves there is special affection for the little garland of putter's songs, characteristically describing a most disagreeable and exasperating job in terms of rueful humour.

The Colliers' Rant

As me an' me marra was gannin' te wark,
We met wi' the devil, it was in the dark.
Aa up wi' me pick, it was in the neet.
Aa knocked off his horns, likewise his club feet.
Chorus: Follow the horses, Johnny me laddie.
Follow the horses, Johnny lad, o.
Follow the horses, Johnny me laddie.
Hey lad, lie away, me canny lad, o.

As me an' me marra was loadin' the tram,
The leet it went oot an' me marra went wrang.
Ye wad ha' laughed at such a fine gam;
Old Nick took me marra but Aa got the tram.
Chorus:

O marra, O marra, oh, what does tha think?
Aa've broken me bottle an' spilt aal me drink.
Aa've lost aal me tools amang the big styens.
Draw me te the shaft, it's time te gan hyem.
Chorus:

O marra, O marra, oh, where hes tha been?
Drivin' the drift from the low seam.
Drivin' the drift from the low seam.
Haad up the lowe, lad, de'il stap oot thy een!
Chorus:

Noo, there is me horse an' here is the tram.
Two tins full o' grease will make her te gan.
There is me marra stretched oot on the groond.
Ye can tear up his shirt, for his minin's aal done.
Chorus:

haad up the lowe = hold up the light

Call the Horse, Marrow

Call the horse, marrow,
For I can call nane.
The heart of my belly
Is hard as a stane.
As hard as a stane
And as round as a cup.
Call the horse, marrow,
Till my hewer comes up.

Me an my marrow
And Christy Crawhall
Will play any three in the pit
At the football.
At the football
And at the coal-tram,
We'll play any three in the pit
For twelve-pence a gam.

Hewing and putting
And keeping in the sticks,
I never (so) laboured
Since I took the picks.
I'm going to my hewer's house
On the Fell Side.
He hews his coals thick
And drives his bords wide.

The rope and the roll
And the long ower-tree,
The devil's flown o'er the heap
With them all three.
The roll hangs across the shaft;
De'il but it fall
And stick in the thill,
Twenty-four horned owls
Run away with the mill.

I'm going to my hewer
Wherever he be.
He's hipt of a buddock
And blind of an e'e.
He's blind of an e'e
And lame of a leg.
My uncle Jack Fenwick
He kissed my aunt Peg.

call = drive

Jacky in the Law Raw

Jacky in the Law Raw,
Canny hinny, bonny hinny,
Jacky in the Law Raw,
Canny hinny now.

My laddie hews twenty every day,
Canny hinny, bonny hinny,
Headways and bordways, headways away,
Canny hinny now.
My laddie hews twenty every day,
Canny hinny, bonny hinny,
Headways and bordways, headways away,
Canny hinny now.

Jacky in the Law Raw,
Canny hinny, bonny hinny,
Jacky in the Law Raw,
Canny hinny now.

My laddie is a kerving hewer,
Canny hinny, bonny hinny,
He hews twenty every hour,
Canny hinny now.
My laddie is a kerving hewer,
Canny hinny, bonny hinny,
He hews twenty every hour,
Canny hinny now.

Jacky in the Law Raw,
Canny hinny, bonny hinny,
Jacky in the Law Raw,
Canny hinny now.

The Miner's Life

Come, gentlemen all, and I'll soon let you know
What hardships the miners they do undergo.
There's none but the sailor on earth to be found
Suffers hardships like miners that work underground.

A collier, they say, is a man of his stain,
Is deprived of sweet air and the day seldom seen.
The work is so hard, every nerve is close braced,
Like a man going to fight, he's stripped down to the waist.

If colliers there were none, what must we do, pray,
To wash, brew or bake, or boil water for tea?
But the brave and bold miner, a man of renown,
Will reek like a kiln to get his coal down.

Besides the great dangers going up and going down,
Some are lamed, some are killed, some are cracked on the crown.
When the water breaks in, oh, it's dismal to see,
When death stares him in the face, and drowned he must be.

Then there's stones from the roof, boys, sometimes they will fall
And break leg or arm, so I'm telling you all,
There's slips from the face that will kill him stone dead,
But the brave living miner still holds up his head.

So now to conclude and to finish my song,
I hope you're not weary with it being long.
Long life to all colliers that work underground,
For I know their equals will never be found.

The Collier

I'm a collier, it's true, and I love a full quart,
And I do delight in a merry good heart.
I smoke my pipe and I dance and I sing.
I'm black, but as happy as George our great king.

I'm a lad, to the coal-pit each morning I go.
I ride in the skip to the coal down below.
And with my small pick, if I work very well,
Black work gets white money and gold also.

I strip coat and waistcoat and shirt off, to work,
And he that's a collier he never must lurk.
We must get some money for our good wives,
And they shall have tea all the days of their lives.

Sometimes I sit cross-legg'd when using my pick,
Which makes the sweat run down when hard I do strike,
Whilst up above me the great coals they do crash,
The pick do strike fire, all around me it flash.

In the midst of great dangers we work and we sweat;
The engine above us it draws up the wet.
When Jack says, 'Come up', and when Jemmy says, 'Haul',
The coals shake the earth whenever they fall.

The Collier Lass

My name is Polly Parker, I'm come o'er from Worsley.
My father and my mother work in the coal mine.
Our family's large, we have got seven children,
So I am obliged to work in the same mine.
And as this is my fortune, I know you feel sorry
That in such employment my days I shall pass,
But I keep up my spirits, I sing and look merry
Although I am but a poor collier lass.

By the greatest of dangers each day I'm surrounded.
I hang in the air by a rope or a chain.
The mine may fall in, I may be kill'd or wounded,
May perish by damp or the fire of the train.
And what would you do were it not for our labour?
In wretched starvation your days you would pass,
While we could provide you with life's greatest blessing.
Then do not despise the poor collier lass.

All the long day you may say we are buried,
Deprived of the light and the warmth of the sun.
And often at night from our beds we are hurried;
The water is in, and barefooted we run.
And though we go ragged, and black are our faces,
As kind and as free as the best we'll be found,
And our hearts are as white as your lords in fine places
Although we're poor colliers that work underground.

I am now growing up fast somehow or other.
There's a collier lad strangely runs in my mind,
And in spite of the talking of father and mother
I think I should marry if he was inclined.
But should he prove surly and will not befriend me,
Another and better chance may come to pass,
And my friends here I know, to him will recommend me,
And I'll be no longer a collier lass.

The Bold Miner

It's of a bold miner who works underground,
He's sweating and toiling his every day round.
He strains and he pines his poor life away,
While the others lie snug in their beds until day.

So when at the bottom in the cage he does land,
He'll hurry inbye with his lamp in his hand,
He'll see all is safe and no gas in the way,
Before he does settle to his work for the day.

Then from the flatt he takes in a tub,
Sometimes it happens he's in a lowse jud,
The coal is so canny, he gets a good day,
To make up his little bit wage at the pay.

But sometimes it happens he's in a hard wall.
The coal is so hard he can hew nyen at all.
He goes to Australia because work is slack,
And, my bonny lads, he will never come back.

Down in a Coal-Mine (I)

In me you see a collier, a simple honest man
Who strives to do his very best to help his fellow man.
We toil away from morn till night where hard work's to be found,
Digging dusty diamonds from underneath the ground.
    Down in a coal mine underneath the ground,
    Where a gleam of sunshine never can be found,
    Digging dusty diamonds all the year round
    Down in a coal mine underneath the ground.

In the morning when we go to toil, it's down the mine we go,
Contented with our lot in life and free from care or woe.
We often think of home and wife and hearts that's filled with mirth
While digging up the fuel from the bowels of the earth.

You often read of accidents that happen down the mine,
How hundreds of poor colliers are shortened of their time;
Explosions they are numerous, and caused by fire-damps
Which when the gas escapes it comes in contact with our lamps.

But when the winter comes in, the collier's worth is found.
Old England's commerce it is spread to all the nations round.
Let those at home rejoice and sing with heart and voices full,
For what would England do without the boys that dig for coal?

Down in a Coal-Mine (II)

I am a jovial collier, as jovial as can be,
But if the trade is very bad it means a lot to me.
And if I stumble with my tongue, I've one excuse to say,
For it's not the collier's heart that's wrong, it's the head that goes astray.
    Down in a coal-mine underneath the ground,
    Where a gleam of sunshine is never to be found,
    Digging dusky diamonds all the season round,
    Down in the coal-mine underneath the ground.

How bravely all them collier lads they toil beneath the ground,
Digging for the coal as the days and nights goes round,
And anxiously their fam'lies wait; how often it is said:
You never know by nightfall how many may be dead.
    Down in a coal-mine underneath the ground,
    Where a gleam of sunshine is never to be found,
    Digging dusky diamonds all the season round,
    Down in a coal-mine underneath the ground.

How little do the rich men care, who sit at home secure,
What dangers all the colliers dare and hardships they endure.
The very fires they have at home to cheer them and their wives
Perhaps were kindled at the cost of jovial colliers' lives.
    Down in a coal-mine underneath the ground,
    Where a gleam of sunshine is never to be found,
    Digging dusky diamonds all the season round,
    Down in a coal-mine underneath the ground.

Five in the Morning

At five in the morning the miner doth rise,
   To go to his dangerous labour.
His bottle and bag round his neck he doth hang,
   And calls on his next door neighbour.

His wife and his children he kiss'd them in bed,
   And bade them a hearty good morning;
Be careful, dear husband, the wife often said,
   And what time will you be returning?

There, look to our children, the miner replied,
   The hour is uncertain for returning.
To work I must go to earn you some bread.
   I may stay in the pit till the morning.

With their picks, drills and scrapers, they descend the deep shaft,
   With their lamps dimly burning,
Through the dark bowels of the earth, and the place of his berth,
   And to wait for the fireman returning.

All right, said the fireman, the miner goes in,
   And strips off his clothes to his breeches.
The pick he doth swing, and the miner doth sing,
   As he shoots the black coal down in pieces.

Then a hole he will drill, some shot he'll put in
   Which shoots a thrill into his neighbour,
But get pure air into the place for his berth,
   And make ease to the poor miner's labour.

At seven in the morning the work they begin,
   The miners are soon in full working,
Some picking, some drilling, some blasting the rocks,
   And others the waggons are filling.

The jig flies with speed, down the steep road
  The horse to the shaft it will take them.
They're put in the cage, and with a loud knock
  A lift with the engine will take them.

Then look to the danger the miner is in,
  As he hangs in the air night and morning.
The rope it may break, he in pieces be dashed,
  And his wife left at home in grief mourning.

An accident of gas sometimes will take place,
  That gives the poor miner no warning;
A fall from the roof sometimes causes death,
  And his children his absence be mourning.

Come all you coal miners who hear these few lines,
  Come look to the old and the feeble.
The wife and the children you know they'll want bread,
  And to work you know they're not able.

Some have dimbly limbs, some to darkness are doomed;
  Then stretch forth your hand like a brother,
Through the dark bowels of the earth where thousands meet death,
  It's our duty to help one another.

Collier's Song

In the depth of a coal-pit a young lad grew,
  The pride of his mother's care,
All round his waist the coal waggon belt he drew,
  And his shoulders were stripped quite bare;
Thus he toiled the day long as he crept through the bay,
  And chained to his load like an ass,
Though hard was his work yet too little was his pay,
  In youthful days to pass.
    Then how long will the colliers and waggoners stand
    The real black slaves o' their native land?

How many are found either buried or drowned,
  And all the whole are oft away swept,
Or they perish by damp or the fire of a train,
  Which brings to an awful death.
Then the mother may weep for her only son,
  And the wife for her husband dear.
Whilst our families starve in a pauperish home,
  And how little do our masters care.
    Then how long will the colliers and waggoners stand
    The real black slaves o' their native land?

The banksman at top he told us he knew,
  That union made us strong,
That each man to another should be true,
  And I echo these words in song.
Then let us unite to seek for our rights,
  And like Britons go hand in hand,
We deserve to be paid for the labour we have made
  In the bowels of our own native land.
    Then how long will the colliers and waggoners stand
    The real black slaves o' their native land?

At five in the morn we leave our home,
  Our wives and children dear,
And perhaps we may never return,
  To our native homes so dear.
But all the long day we are deprived of light,
  And of the warmth of the sun.
And whilst we are in rags in poverty we run,
  And this is our native land.

The Collier Lad is a Canny Lad

Oh, the collier lad is a canny lad,
An' he's aalwes of good cheor,
An' he knaas how te wark, an' he knaas how te shork,
An' he knaas how to sup good beor.

Wey, it's doon the shift on a Monday morn,
An' the kevil is the best,
In the Busty seam wi' Thompson's team,
In a flat called the Fowerteenth West.
An' the fyece is a hundred an' five yards lang
When measured fre neuk to neuk,
An' when craalin alang the scufflins, lads,
Keep doon or ye're boond te get stuck.

When the shots gan off, then the shovels do fly
Till the belt gets loaded full,
An' in half an oor a stone gans on,
An' the motor will not pull.
'Brokken belt!' is the cry, an' we aal creep oot,
Te the mothergate it te mend.
Geordie Hall, he's the depitty in wor flat,
Says: 'Ye'll drive us roond the bend.'

So we pull an' we strain for te fix it agyen,
An' when it's been put straight,
Tim Jones, that's the seccetary of wor Lodge,
Says: 'It's time that ye had your bait.'
So we tak worsels te a quiet spot,
Wi' a plank an' a chock for a seat,
An' the crack at last flies thick an' fast,
O' the deeins at the Club last neet.

Wey, it's very hard when ye're paid by the yard
For te tak lang ower your bait.
So we craal back on, get some timmerin done,
For the belts we can hardly wait.
For it's twenty-six inches high, me lads,
An' the wark is really grand.
An' the fillers' pay, fower quid a day,
It's the best in aal the land.

> Oh, the collier lad is a canny lad,
> An' he's aalwes of good cheor,
> An' he knaas how te work, an' he knaas how te shork,
> An' he knaas how te sup good beor.

The Fore Shift

Oh, the fore shift dark and dreary,
Oh, this lonely two o'clock;
Limbs may ache, and hearts be weary,
Still there comes the caller's knock;
And each blow upon the panels
Bids us up and don our flannels,
By the light of lamp or can'les
Batter at the grimy rock.

Just to get a bare subsistence,
Little earned and nothing saved,
With the workhouse in the distance
After we for years have slaved.
Some look on with holy horror,
At each pitman's little error,
But 'twould much abate their terror
Could they see the dangers braved.

To the coal's grim face we travel,
And again our flannels doff;
Can they wonder if we cavil
At the ones much better off?
Like a snake our bodies coiling,
Weary hours of incessant toiling,
Through each pore the sweat comes boiling;
Think on this, ye ones that scoff!

Up while stars are dimly peeping
Through the midnight's sable gloom;
Up while pampered ones are sleeping
In their snug and cosy room.
Fore shift visions need not haunt them,
Nor the pit's grim danger daunt them;
Oh, 'twas kind of fate to plant them
Where they could so safely bloom!

The Haswell Cages

Come all you good people and listen a while.
I'll sing you a song that will cause you to smile.
It is about Haswell I mean for to sing,
Concerning the new plan we started last spring.
    And the very first thing I will mention,
    Without any evil intention,
    It's concerning this new invention
    Of winding up coals in a cage.

It was in eighteen hundred and thirty eight,
We begin to prepare to make the shaft right.
We put in the conductors from bottom to top,
The materials were ready prepared at the shop.
    From the top of the pit to the bottom:
    One hundred and fifty-six fathom.
    And the distance you do think it nothing,
    You ride so quickly in the cage.

Now, considering the depth, it's surprising to say
The quantity of work we can draw in a day.
Five hundred and thirty tons of the best coal
In the space of twelve hours we can win up this hole.
    About forty-five tons in an hour.
    And, viewers, overmen and hewers,
    Our engines must have a great power
    To run at such speed with the cage.

Then as soon as the tubs do come to the day,
To the weighing machine they are taken away,
Where two men are appointed there to attend
To see justice done between master and men.
    And when they leave the weighing machine, sir,
    Straightway they do go to the screen, sir,
    And the keeker does see that they're clean, sir,
    All the coals that come up in the cage.

I've wrought with the corves, I have wrought with the tubs,
I have wrought where the baskets come up by the lugs,
I have wrought by the dozen, I've wrought by the score,
But this curious contrivance I ne'er saw before.
    When we get in, they then pull the rapper.
    At the top it does make a great clatter,
    And the brakesmen they know what's the matter,
    And bring us away in a cage.

And when the bell rings and the top we approach,
It oft puts me in mind of a new railway coach.
The number of passengers I cannot tell,
But she brings a great many, I know very well.

But I wish they may not overload her,
And do some mischief on the road, sir.
Too much charge makes a cannon explode, sir,
And so will too much in a cage.

Now the young men and maids sometimes take a trip
Out to sea in fine weather aboard a steam-ship,
But if any be curious enough to engage
For a trip down below and a ride in our cage,
It would be a fine recreation
For to go down and view the low station.
I wish they may meet no temptation
When they take a trip in our cage.

The Winder

I sit on me chair and I watches me gauge,
I let off me brake and I lower the cage,
And God, am I careful, for on that one rope,
Sixty-four colliers place all their hope.

Me engines is steam and I like it that way,
Though I know the electric's the thing for today,
But I know me old steamers and I think they know me,
These were built in Kilmarnock in nineteen-o-three.

I keep 'em well oiled and in perfect trim.
I rub 'em wi' polish on a piece of old scrim,
For I like to be sure that they'll do their work well,
That they'll pull 'er up smooth, when Old Bob rings the bell.

I don't mind the noise, there's enough of it though,
An' I talks to me engines, and I think as they know,
Because when I tell 'em there's men up from work,
They run sweet as a nut, an' wi' never a jerk.

When we're windin' the coal, I just give 'em their head.
Couldn't do that wi' men, why, we'd have 'em all dead!,
Cause when they're goin' flat out, they do go at some pace,
Sometimes I'm sure as they're trying to race!

Aye, I like me own job; not much money it's true,
But what else could a lazy old stick like me do?
So I'll just sit on me chair an' I'll study me gauge,
An' all as I ask is a fair livin' wage.

The Cage

Monday morning, Shift Number One.
Wish that the day's work was over and done.
Climb in the cage, boys, let her roll!
We're the warriors who fight for the coal.
    The wheels of the headgear spin around.
    The cage like a stone is droppin' down.
    Feel the blood in your ears pound.
    Down, down into the ground.

First stop bottom, fall like a stone.
Halfway down, lads, soon be done.
There's a hell of a draught, a howlin' gale,
As the cage whistles downwards, a-tellin' its tale.

Glarin' lights, dusty floor,
Meet us there outside the cage door.
Crawl to the face, lads, sharin' a joke.
In a couple of hours you'd die for a smoke.

Sweat runs freely, matted hair,
Glistenin' bodies, black and bare.
It's damned hard work but what d'you say?
Roll on Friday when we draw our pay!
    The wheels of the headgear spin around.
    The cage like a stone is droppin' down.
    Feel the blood in your ears pound.
    Down, down into the ground.

The Depth o' the Mine

Aal aboard for the bottom, the brakesman wad yell,
We'd step in the cage an' he'd ding-dong the bell,
Through that hole i' the Moor, like a hooer into hell
We'd drop doon to the depth o' the mine.

In wet or dry weather, the shaft always rains,
Wi' a tremblin o' timber an' clankin o' chains.
Aw wis just off a spree. Gox, it shook me poor brains,
Droppin doon to the depth o' the mine.

The Driver

There was ne sign o' Schuel Boards when Aa was a lad.
Aa went inte the pit when but nine summers aad,
To keep a trap door on an aad rolley way,
For the wonderful wage of a shillin a day.
But aalways bein wishful te cut an' contrive,
Te better me wages Aa started te drive;
An' oh, ye should 'a' heard how Aa whustled an' sang
As Aa rattled the full 'uns an' chum 'uns alang.
    Siss, siss! Baa, baa!
    That's hoo we drive them doon belaa.

Aa remember the very first morning Aa yoked,
Hoo the landin lads laughed an' the putters aal joked.
But a shift or two myed me se used to the wark,
Aa could yoke 'em or lowse 'em or drive in the dark.
For Aa knew ev'ry curve, ev'ry way-end an' switch,
Where te grape for the dregs when Aa come tiv a hitch,
Hoo te splice a good whup, hoo te use a one tee!
Eh, we did not do badly, aad Smiler an' me.
    Siss, siss! Baa, baa!
    That's hoo we drive them doon belaa.

Aa confess that Aa liked a bit fun te create,
An' Aa've jolloped the drivers that's stolen me bait,
Had me hand in at fondies a' aal such as that,
But Aa'd never let putters wait-on at the flat.
When sets would get off, or the horses would kick,
An' the drivers would promise the weight of a stick,
Aa could aalways contrive te keep well oot the way,
When the owerman ettled the piper te pay.
    Siss, siss! Baa, baa!
    That's hoo we drive them doon belaa.

When the men had ne coals or there'd been a big fall,
Aa've swapped midgies, Aa played 'Droppy candle' an' aal,
An' wi' lads like meself, when nowt else could be deun,
Aa've tossed up me pit cap an' Aa've fitten for fun.
But it's useless for me to reflect on the past –
The bit pluck Aa had then'll be mine till the last,
An' whatever God sends us, Aa'm willin' te bear,
But Aa'm certain that Aa'll never drive ony mair.
Siss, siss! Baa, baa!
That's hoo we drive them doon belaa.

grape = grope
ettle = intend
fitten = fought

I am a Driver

I am a driver, these are my tubs.
I'm off the road, boys, and my pony rubs.
Where is the doggy? Nobody knows.
He'll be out on the pass-by, a-pickin' his nose.

I shall be glad when this shift is done,
Then I'll be up there out in the sun.
Tha'll be dahn 'ere, boy, in this dark 'oil,
Still gruntin' and groanin', pullin' this coil.

Corn's in the manger, water's in t'trough.
Tha'll shove thee nose out when tha's 'ad enough.
I'll tek thee in t'standin' and drop off thee gears.
When I come back I know tha'll be 'ere.

The Stoneman

Now the coal is off an' the fillers hev left the flat,
An' the Deppitty comes an' asks us what we're at,
So we tell him, an' he stems the holes we've drilled,
An' the shots gan off an' the packs we build
Doon the Brockwell Seam in the north of No. 5 West.

Now we build packs from the stones that's in the gate,
An' we get them neat an' tidy weel afore bait,
For there's gorders te hump an' gorders te set,
An' side te take off, that aa dor bet,
Doon the Brockwell seam in the north of No 5 West.

So the belt is rolled an' set down on the new tracks,
An' they draa oot the chocks while lyin' on their backs,
An' they set them again in thor new place,
For te stop the weightin' on the face
Doon the Brockwell Seam in the north of No 5 West.

Well, aa'm a' reet the morn for a sup, aa knaa,
For me marra's reckoned up that he's won the snaaball draa,
Wi a hoosy card an' a pint in me hand,
Aa'll forget aal aboot that No-man's-land
Doon the Brockwell Seam in the north of No 5 West.

snaaball draa = snowball draw, club lottery
hoosy card = Bingo card ('housey-housey' is the army name for Bingo)

The Pick-Sharpener

Bring your blades to me, me boys, and I'll give 'em all an edge.
Bring your shovels for reshaftin, bring your ringer or your wedge.
  The pick-sharpener they call me, boys, an' I'm happy at me job,
  Wi' me furnace an' me anvil just to make an honest bob.

Send your picks up on the cage, put your numbers on wi' chalk,
An' I'll send 'em down all sharpened just to save you lads the walk.
You can pay for 'em on Friday, or else pay me when you can,
For I know the Kirkby collier, an' he's an honest man.

When the gaffer comes around, he shouts: 'How you doing, Jim?',
Since he doesn't mind a joke, I say I'm not so well as him.
I tell him the work's too hard an' the money's not enough,
Then he laughs an' says: 'You spend it on that wind an' water stuff.'

No, I wouldn't swap me job, me boys, for fifty quid a week.
Every collier knows me, and everyone will always speak.
When they see me in the pub, me boys, they'll buy me many a pot,
So when I get to work next day, me sharpenin's not so hot.
  The pick-sharpener they call me, boys, an' I'm happy at me job,
  Wi' me furnace an' me anvil just to make an honest bob.

Oakey's Keeker

O Oakey's! O Oakey's! What makes you so bad?
It's enough for to make all your workmen go mad.
We should like very much to know what you mean
The way you go on from the pit to the screen.
You treat us coal hewers just as you think fit.
The wages are small that are paid in the pit;
But what we are making we really don't know,
Since they have sent us old Maiden Law Joe.

This famous old keeker must not understand
How we are tormented with ramble and band.
The ramble comes down, after firing a shot,
Among the loose coals, and it cannot be got.
By the light of a candle it cannot be found;
Daylight is different from being underground.
If this old keeker would only think so,
We would speak better of old Maiden Law Joe.

To do his duty is nothing but right,
But in hurting coal-hewers he takes a delight.
If he pleases the masters, that's all he cares for,
Suppose that he hungers poor men to the door.
At half-past six in the morning he starts
To fill up the box, which is only two quarts;
If he gets the first tub, how pleased he will go,
And say: 'That's a start for old Maiden Law Joe.'

He was at the Bank Foot – that's near to the Plain.
We wish he was only back there again.
While he was there he was doing the same.
He must have been born without feeling or shame.
They say there's a medium in every case;
He's not a fit man to have such a place,
For he has no feeling for men that's below,
This hairy-faced rascal, old Maiden Law Joe.

This Maiden Law tyrant does nothing but shout:
'Who belongs to this tub? Because it's laid out.'
He smacks his old lips, his hands he will rub,
Because he has taken a good man's tub.
Amongst us coal-hewers how well he is known;
His hardness towards us he always has shown.
What makes him do it, I really don't know –
That cruel impostor, old Maiden Law Joe.

I hope all the screeners, as well as old Joe,
Will think of the men that are working below.
Perhaps in a pit they may never have been.
That's where the hardship may daily be seen.
How would they like it, if they knew what we made,
When the Pay came, and the money not paid?
I hope that the whiskers so quickly will grow
As to fill up the mouth of old Maiden Law Joe.

Now, Joey Badun, you silly old man,
You have nearly done all the ill that you can.
With age your whiskers are turning quite grey.
I think it is time you were starting to pray.
I never did like to wish anyone harm,
But I doubt you will go to a place where it's warm.
It's nothing but right to reap what you sow,
And they'll burn your whiskers, old Maiden Law Joe.

Now if you take ill, and be confined to bed,
Do you think your masters will keep you with bread?
Do you think your masters will visit you then,
For all you have always imposed on the men?
For all you did for them their money to save,
If you were dead, would they go to your grave?
Not one single step with your corpse will they go,
Because it's that rascal, old Maiden Law Joe.

Now, Mr Badun, I'm writing too long.
I hope you'll forgive if there's anything wrong.
When God calls upon you, what will you say?
Those tubs will be standing before you that day.
If the old Devil sees you, he'll give a great shout –
'That's Oakey's old keeker who laid the tubs out!'
God will then say: 'Down to hell you must go,
If you are the keeker called Maiden Law Joe.'

The Putter

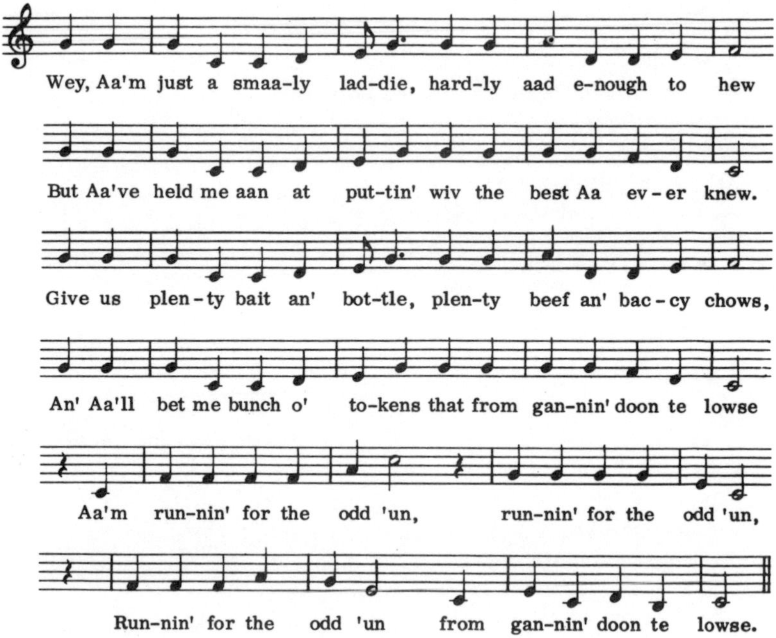

Wey, Aa'm just a smaaly laddie, hardly aad enough to hew,
But Aa've held me aan at puttin' wiv the best Aa ever knew.
Give us plenty bait an' bottle, plenty beef an' baccy chows,
An' Aa'll bet me bunch o' tokens that from gannin' doon te lowse
    Aa'm runnin' for the odd 'un, runnin' for the odd 'un,
    Runnin' for the odd 'un from gannin' doon te lowse.

There's half a dozen gannins in the flat that Aa'm at noo,
An' if Aa'd me aan in choosin' Aa wad hew in one or two;
But somehoo or other, hoo it comes Aa diven't knaa,
As sure as rub me cavil, it's the worst 'un on the saa.
    So Aa'm runnin' for the odd 'un, runnin' for the odd 'un,
    Runnin' for the odd 'un from gannin' doon te lowse.

There's a hitch an' then a swally, filled wi' watter like a ford,
An' a lot o' way aal twisted in the clarty gannin' board.
There's law planks an' raggy canches where you sometimes get a smack,
An' it makes ye twist yer gizzard if ye chance te catch yer back
When you're runnin' for the odd 'un, runnin' for the odd 'un,
Runnin' for the odd 'un from gannin' doon te lowse.

There's a short plate an' a lang 'un near the double torn inbye.
They've been fettled wiv a closer, an' the closer winnet lie.
Oh, the short plate an' the lang 'un, could they speak, what wad they say?
For Aa've telled them lots o' secrets, when Aa've tummelled off the way
As Aa'm runnin' for the odd 'un, runnin' for the odd 'un,
Runnin' for the odd 'un from gannin' doon te lowse.

There's a lang law heavy pillar inbyeside the canvas door,
Where Aa hoss the scrubbin' full 'uns up for eighteenpence a score.
Hoo Aa bless that lang law pillar! Hoo that aaful hitch Aa dread,
For it's fearful wark, this pushin' an' shovin' wi' your head
When you're runnin' for the odd 'un, runnin' for the odd 'un,
Runnin' for the odd 'un from gannin' doon te lowse.

There's a mate o' mine, a filler, wears a shirt o' flannel blue,
Which Aa filled away at Christmas, 'cause he wadn't pay his due.
Ye should have heard him swearin', well, he'd bruise at every bat
As he chased us wiv the shot-stick; but Aa licked him te the flat,
Just like runnin' for the odd 'un, runnin' for the odd 'un,
Runnin' for the odd 'un from gannin' doon te lowse.

The Pony Putter

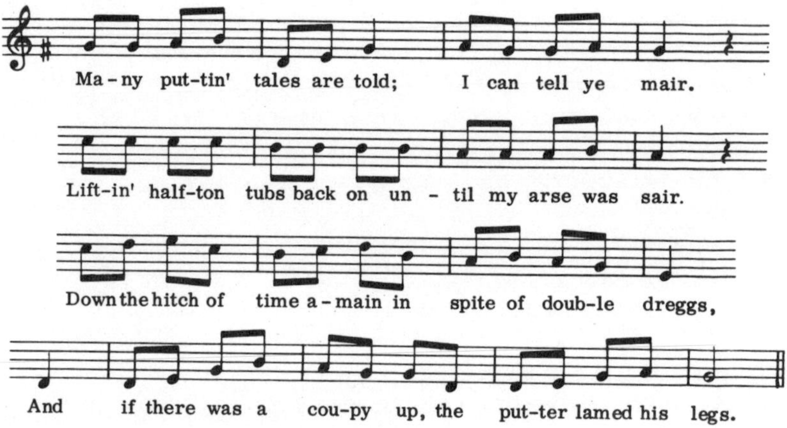

Many puttin' tales are told; I can tell ye mair.
Liftin' half-ton tubs back on until my arse was sair.
Down the hitch of time amain in spite of double dreggs,
And if there was a coupy-up, the putter lamed his legs.

Puttin' was an art, now lost; putters are no more.
The work, I know, was killy-pig for ten and ten a score.
You'll find no pony putters in pits from Tees to Tyne.
I'll tell ye when they went on strike because the sun did shine.

At Pelaw Pit on the heap one sunny summer's day,
Trouble over tokens nearly turned the weighman grey.
A bunch of tokens up they tossed; if they came down again,
Pelaw putters were on strike – so all the lads went hyem.

Waitin' on for chummins, the crack heard at the kist.
The darin' deeds of putters, they didna give a jist.
For men or mine or management, he tret them all the same,
The joker on the limmers; he knew well the puttin' game.

One day on the carpet, the black book we did sign
Because we rode the gallowas; five bob was the fine.
'Five bob, sir,' the putter said; he made the gaffer gape.
'They're not fined that at Durham when they're on a charge o' rape.'

Ye can always tell a lad that's put; there's traits he canna hide –
The trick with double tokens, or the odd tub off the side.
But if you are in trouble, you only have to say;
He'll put his shoulder to the tub, and lift you on the way.

The Volunteer Putter

The deputy in the Main coal flat one Saturday morning said:
'Our putters they are aal off work, will someone put instead?'
An anxious look crept ower the face of every hewer there,
As eagerly they watched to see who'd volunteer that day.

An aad man staggered doon the flat, his hoggers were aad and torn.
How strange aad Andra seemed to be, on the limmers that Saturday morn.
And as his bare legs touched the tub, withoot a single pause,
The language that followed was the warst Aa've ever heard.

> The scene was one Aa'll niver forget as lang as Aa may live,
> An' just to see it ower again, my slush cap Aa wad give.
> The drivers they were aal amazed, the deputy aad and grey,
> Not one of them will ever forget that shift last Saturday.

Aad Andra swore after that day, he'd niver put no more,
But the deputy said he'd done very well, for he had put two score.
Aad Andra says: 'Thou canna kid me, for Aa knaa what Aa'm feelin' like.
If thou wants a putter next Saturday, thou'll hev to send hyem for the wife.'

And as he dragged his pick and shuel alang the gannin' board,
The language that followed was the warst Aa've ever heard.

> The scene was one Aa'll niver forget as lang as Aa may live,
> An' just te see it ower again, my slush cap Aa wad give.
> The drivers they were aal amazed, the deputy aad and grey,
> Not one of them will ever forget that shift last Saturday.

The Doggie Putter

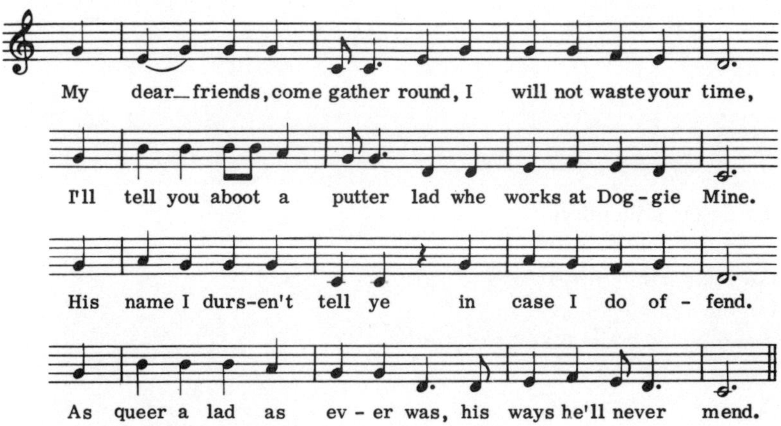

My dear friends, come gather round, I will not waste your time.
I'll tell you aboot a putter lad whe works at Doggie Mine.
His name I dursen't tell ye in case I do offend.
As queer a lad as ever was, his ways he'll never mend.

On Monday morn when he's in forst he gets up feelin' queer,
'Cause just the previous neet afore he'd been oot on the beer.
Amazin pity, grace unknown, an' love beyond degree,
His mother tries te help him, but he's dead as dead can be.

An' then he tries his uppermost te get into his shirt.
He puts his legs doon both the sleeves an' rives it in three parts.
An' next he puts his stockins on, an' finds 'em inside oot,
An' sends a prayer up te his God; the bugger's lost his boots.

At last he manages te get dressed wi' aal his might an' main.
His mother says: 'Good mornin, lad', but his mind is far away.
An' as he toddles doon the yard wi' a rottle in his throttle,
He hoys his push-bike o'er his back an' pedals away on his bottle.

An' as the hours flee in that shift, to him like rollin years,
He struggles on, weary an' worn, wi' the trials that occur.
At last the shift comes te an end. 'Oh, thank the Lord,' he cries,
'Take us te my home at last, an' rest me weary sides!'

in forst = on first shift

Aa wes gannin inbye

Aa wes gannin inbye
On the engine plane.
Aa could hear the putter shoutin':
'Aa'm off the way!
Ah wey, give us a lift,
Me arse is sair.
If Aa had this tub put,
Aa wad put ne mair.
(Ne mair, ne mair!) . . . '

The Putter's Lament

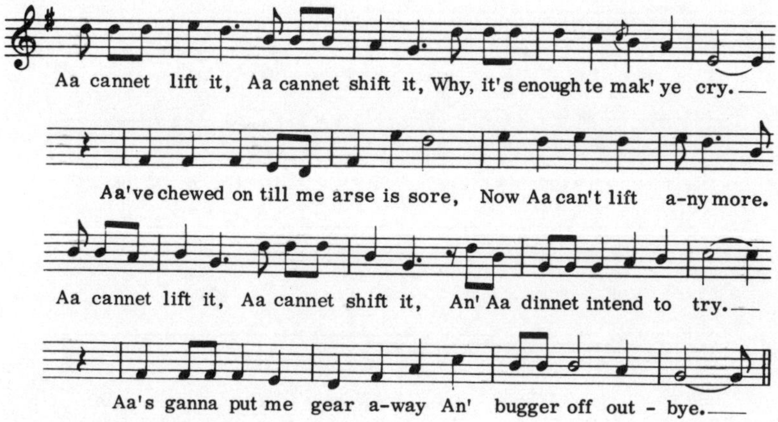

Aa cannet lift it, Aa cannet shift it,
Why, it's enough te mak' ye cry.
Aa've chewed on till me arse is sore,
Now Aa can't lift any more.
Aa cannet lift it, Aa cannet shift it,
An' Aa dinnet intend te try.
Aa's ganna put me gear away,
An' bugger off outbye.

Poor Putter

Wey, he's in, he's out,
He's off, he's on.
He's gettin' his bait,
He's fillin' a one.
There he goes away again.
He's in, he's out,
An' he's off again!

Little Chance (I)

My name is William Robinson,
  That same I do advance.
The first little galloway ever I drove,
  They called him Little Chance.
Chancey had two greasy legs,
  Likewise a kettley back,
And going down the waggon way
  He made the waggons knack.

    So down the pits we'll go, my lads,
      Down the pits we'll go.
    We'll try our will and try our skill
      To putt the ridge below.
    The timber leaders two shillings,
      The deputies half a crown,
    The overman five and sixpence
      For travelling up and down.

I'm going to putt next quarter,
  To make three pounds o' pay,
It's better'n driving Chancey
  For eighteen pence a day.
So come, you jolly driver lads,
  Get on your limmers and gan,
And work for Mr Hindley,
  For he's a gentleman.

Little Chance (II)

Now, ye gan ower the Busty fields te gan doon the pit.
Ye get yer lamp oot, ye gan inbye, an' there ye sit at the kist.
The deputy says: 'Thee place is holed, thoo'll ha' te gan straight on.'
Aa says te him: 'What's the matter wi' me own?' He says: 'She canna gan on.'

Aa got fifteen oot of a jud, titty fa la, titty fa lay.
Eh, by hell, she wes good, titty fa la, titty fa lay.
Aa came oot te get a shaft, when the timmer it gie a crack,
An' a stone fell on me back, titty fa la, titty fa lay.
Tra la lala la la, ower the waal's oot!

Noo, ye're sure te ken me brother Mick, for he's se full o' wit.
He got a job o' puttin' up at 'Cotia pit.
When he comes hyem fre wark, he's like a droonded rat;
Instead o' gannin upstairs te bed, he lies upon the mat.
Bugger, he puts a thoosand or more, titty fa la, titty fa lay,
They pay him by the score, titty fa la, titty fa lay.
He fills his tubs se quick, withoot any delay,
But he can never find his pick, titty fa la, titty fa lay.
Tra la lala la la, ower the waal's oot!

Noo, Jack an' Bill, two marras, were in a public hoose.
The talk aboot the cavils, man, it wadn't frighten a moose.
Jack says te Bill: 'Eh, by God, she's hard!
The tops is like bell-metal but the bottoms is not se bad.
Bugger, Aa only got fower the day, titty fa la, titty fa lay.
Aa only got fower the day, titty fa la, titty fa lay.
Aa'd have getten other fower, Aa wes wishin the shift wes ower,
When the putter got off the way, titty fa la, titty fa lay.'
Tra la lala la la, ower the waal's oot!

Noo, me name is Jackie Elliott, me name Aa do advance.
Aa drive a little gallowa, they caal him Little Chance.
Two greasy feet, likewise a kittly back,
An' gannin alang the gannin-board he makes the chum 'uns knack.
Aa wes gannin aroond the torn, titty fa la, titty fa lay.
Chancy wadn't haad on, titty fa la, titty fa lay.
The tubs they gie a click, Aa got off the way te switch.
Ye bugger, Aa smashed the deputy's kist, titty fa la, titty fa lay.
Tra la lala lala, ower the waal's oot!

Noo, me an' me wife, me mother-in-law, went te the silvery sea.
The mother-in-law gat into a boat, a sailor she wad be.
She hadn't gan passin twenty yards when suddenly there's a shoot;
Me mother-in-law falls into the sea, an' there she's splashin aboot.

She shouts: 'Help, aa canna swim!' Titty fa la, titty fa lay.
Aa says. 'Noo's the time to learn.' Titty fa la, titty fa lay.
Me wife, she says: 'Ye hoond, tha's not ganna watch her droond!'
Aa says: 'Naw, aa'll shut me eyes.' Titty fa la, titty fa lay.
Tra la lala la la, ower the waal's oot!

Jowl an' Listen

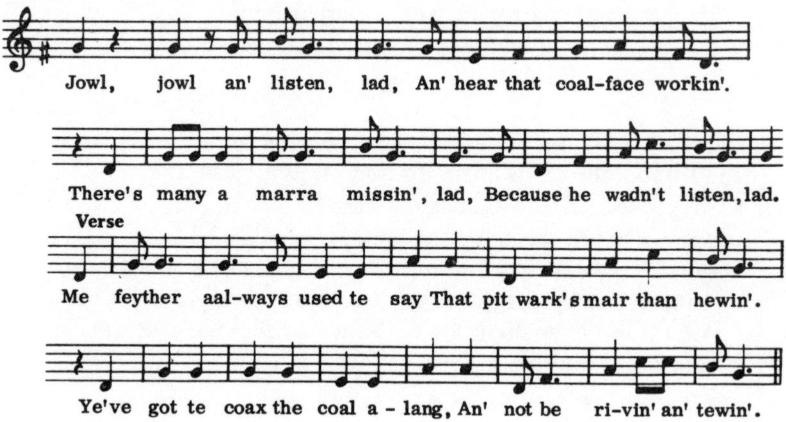

Jowl, jowl an' listen, lad,
An' hear that coal-face workin'.
There's many a marra missin', lad,
Because he wadn't listen, lad.

Me feyther aalways used te say
That pit-wark's mair than hewin'.
Ye've got te coax the coal alang,
An' not be rivin' an' tewin'.

So jowl, jowl an' listen, lad,
An' hear that coal-face workin'.
There's many a marra missin', lad,
Because he wadn't listen, lad.

Noo, the depitty craals fre flat te flat,
While the putter rams the tyum 'uns,
But the man at the face hes te knaa his place
Like a mother knaas her young 'uns.

So jowl, jowl an' listen, lad,
An' hear that coal-face workin',
There's many a marra missin', lad,
Because he wadn't listen, lad.

Early in the Morning

Early in the morning, down, down, down,
They wind us deep in the pits.
We work like niggers, we cut such funny figures,
All among the nobbly bits (Lord love yer!).
We work all day for very little pay,
That's what puts us in a rage.
For we go down white, and back we come at night
Like a lot of little blackbirds in a cage.

II

The Miner at Play

II

THE MINER AT PLAY

Before the Industrial Revolution, what little leisure working people found they filled with flourish. On holidays there were craft pageants and labourers' festivals, with singing and dancing and a gay show of ribbons. The pitmen and their families would gather of an evening on the coaly green and dance to the tunes of the fiddler or piper, who might be a talented workmate or a humble professional who had tramped out of town for the occasion. *Wreckenton Hiring* has the ring of a time when even the roughest classes had a strong instinct for gaiety and beauty, and *The Colliers' Pay Week* gives a picture, as clear, detailed and devoted as a Bewick cut, of the spirited pitmen of Benwell, near Newcastle, at the start of the nineteenth century. It is remarkable how many of these early colliery songs remind us of Thomas Bewick, and indeed the scenes of pay-week were just the kind of thing to delight his sharp eye, when the year's bond was up, and the colliers with primroses in their caps and money in their pockets would tramp into Newcastle with their girls, a laughing, boozing, proud and genial lot, as quick for a dance as for a fight or a kiss.

Early in the nineteenth century the blackness and weight of the Industrial Revolution began to tell. Exploitation, insecurity and ugliness withered the spirit of millions of working people, and left them spare and sullen, without the heart to sing. But in the coalfields the obstinate balladeers remained, no longer celebrating a dance on the green, but still bawling of race-horses and fighting-cocks, of practical jokes down in the pit and of uproar among the pub fiddlers and the women of the town, and of the young miners zigzagging their way home after a pay-Friday booze-up, as pictured in the modern song called *Stottin' doon the Waall.* Some songs of the big regional festivities and galas appear in a later part of this collection, but the handful of samples offered here remind us that to give only a dark picture of miners' life is as false as to present it as idyll.

Walker Pits

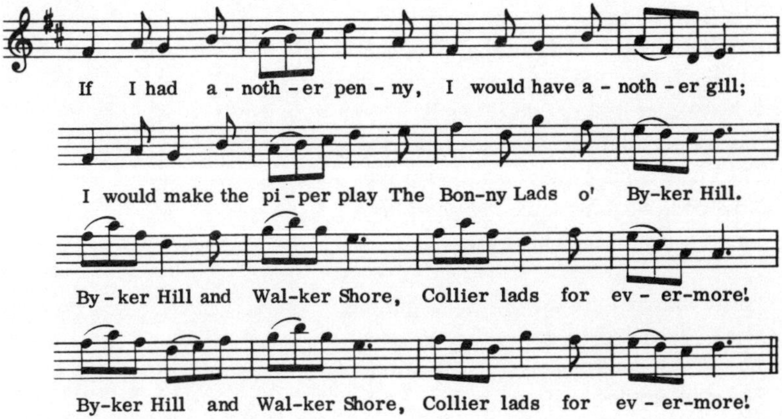

If I had another penny
I would have another gill.
I would make the piper play
The Bonny Lads o' Byker Hill.

> Byker Hill and Walker Shore,
> Collier lads for evermore!
> Byker Hill and Walker Shore,
> Collier lads for evermore!

When I came to Walker wark,
I had ne coat nor ne pit sark;
But now I've getten two or three.
Walker Pit's done well for me.

> Byker Hill and Walker Shore,
> Collier lads for evermore!
> Byker Hill and Walker Shore,
> Collier lads for evermore!

The Colliers' Pay Week

The Baff Week is o'er – no repining –
Pay-Saturday's swift on the wing;
At length the blithe morning comes shining,
When kelter makes colliers sing.
'Tis spring, and the weather is cheery,
The birds whistle sweet on the spray;
Now coal-working lads, trim and airy,
To Newcastle town hie away.

Those married jog on with their hinnies,
Their canny bairns go by their side;
The daughters keep teasing their minnies
For new clothes to keep up their pride.
They plead Easter Sunday does fear them,
For if they have nothing that's new,
The crow, spiteful bird, will besmear them.
Oh, then what a sight for to view!

The young men, full blithesome and jolly,
March forward, all decently clad;
Some lilting up *Cut-and-Dry Dolly*,
Some singing *The Bonny Pit Lad.*
The pranks that were played at last binding
Engage some in humorous chat.
Some halt by the wayside on finding
Primroses to place in their hat.

Bob Cranky, Jack Hogg and Dick Marley,
Bill Hewitt, Luke Carr and Tom Brown,
In one jolly squad set off early
From Benwell to Newcastle Town.
Such hewers as they (none need count it)
Ne'er handled a shovel or pick.
In high or low seam they could suit it,
In regions next door to Old Nick.

Some went to buy hats and new jackets,
And others to see a bit fun;
And some wanted leather and tackets
To cobble their canny pit shoon.
Save the ribbon Dick's dear had requested
– Aware he had plenty of chink –
There was no other care him infested,
Unless 'twere his care for good drink.

At length in Newcastle they centre,
In Hardy's, a place much renowned,
The jovial company enter
Where stores of good liquor abound.
As quick as the servants could fill it
– Till emptied were quarts half a score –
With heart-burning thirst down they swill it,
And thump on the table for more.

With boozing and laughing and smoking,
The time slippeth swiftly away;
And while they are ranting and joking,
The church clock proclaims it mid-day.
And now for black puddings, long measure,
They go to Tib Trollibag's stand,
And away bear the glossy rich treasure
With joy, like curled bugles in hand.

And now a choice home they agreed on,
Not far from the head of the Quay,
Where they their black puddings might feed on,
And spend the remains of the day,
Where pipers and fiddlers resorted
To pick up the straggling pence,
And where the pit lads often sported
Their money at fiddle and dance.

Blind Willie the fiddler sat scraping
In a corner just as they went in.
Some Willington callants were shaking
Their feet to his musical din.

Jack vowed he would have some fine capering,
As soon as their dinner was o'er,
With the lassie that wore the white apron,
Now reeling about on the floor.

Their hungry stomachs being eased,
And gullets well cleared with a glass,
Jack rose from the table and seized
The hand of the frolicsome lass.
'Maw hinny,' says he, 'pray excuse me –
To ask thee to dance aw make free.'
She replied: 'Aw'd be loth to refuse thee.
Now fiddler, play *Jigging for me*.'

The damsel displays all her graces,
The collier exerts all his power;
They caper in circling paces,
And set at each end of the floor.
He jumps, and his heels knack and rattle,
At turns of the music so sweet.
He makes such a thundering brattle,
The floor seems afraid of his feet.

This couple being seated, Bob rose up.
He wished to make one in a jig;
But a Willington lad set his gob up –
O'er him there should none 'run the rig'.
For now 'twas his turn for a caper,
And he would dance first as he'd rose.
Bob's passion began for to vapour;
He twisted his opponent's nose.

The Willington lads, for their Frankie,
Jumped up to revenge the foul deed;
And those in behalf of Bob Cranky
Sprung forward, for now there was need.
Bob canted the form with his kevel,
As he was exerting his strength;
But he got on the lug such a nevel
That down he came, all his long length.

All hand-over-hand, topsy-turvy,
They struck with fists, elbows and feet.
A Willington callant called Gurvy
Was top-tails tossed over the seat.
Luke Carr had one eye closed entire,
And what is a serio-farce,
Poor Robin was cast on the fire,
His breeks torn and burnt off his arse.

Oh, Robin, what argued thy speeches?
Disaster now makes thee quite mum.
Thy wit could not save the good breeches
That mensefully covered thy bum.
To some slop-shop now thou may go trudging,
And lug out some squandering coins;
For now 'tis too late to be grudging:
Thou cannot go home with bare groins.

How the wayfaring companies parted
The Muse chooseth not to proclaim;
But 'tis thought that, being rather downhearted,
They quietly went toddling hame.
Now, ye collier callants se clever,
Residing 'tween Tyne and the Wear,
Beware, when you fuddle together,
Of making too free with strong beer.

Wreckenton Hiring

Oh, lads an' lasses, hither come
To Wreckenton to see the fun,
An' mind ye bring yer Sunday shoon,
There'll be rare wark wi' dancin, O
    An' lasses now, wi'oot a brag,
    Bring pockets like a fiddle-bag,
    Ye'll get them crammed wi' mony a wag
    O' pepper-kyek an' scranchim, O.

An' Bess, put on that bonny goon
Thy mother bought thou at the toon,
That straw hat wi' the ribbons broon,
They'll a' be bussed that's comin, O.
    Put that reed ribbon round thy waist,
    It myeks thou luik sae full o'grace,
    Then up the lonnen come in haste.
    They'll think thou's comed frae Lunnen, O.

Ned put on his Sunday's coat,
His hat an' breeches cost a note,
Wi' a new stiffener roond his throat,
He luikt the very dandy, O.
    He thought that he was gaun to choke,
    For he'd to gyep before he spoke.
    He met Bess at the Royal Oak;
    They had baith yell an' brandy, O.

Each lad was there wi' his sweetheart,
An' a' was ready for a start,
When in come Jack wi' Fanny Smart,
An' brought a merry scraper, O.
    Then Ned jumped up upon his feet,
    An' on the table myed a seat,
    Then bounced the fiddler up a beat,
    Sayin: 'Play, an' we will caper, O.'

Now Ned an' Bess led off the ball.
'Play *Smash the Windows*,' he did call.
'Keep in yor feet,' says Hitchy Mall,
'Learn'd dancers hae sic prancin, O.'
  Now Ned was nowther laith nor lyem,
  An' faith, he had baith bouk an' byen.
  Ye wad thought his feet was myed o'styen,
  He gave sic thuds wi' dancin, O.

Now Jacky Fanny's hand did seize,
Cryin: 'Fiddler, tune yor strings to please.
Play *Kiss her weel amang the trees*,
She is my darlin, bliss her, O!'
  Then off they set wi' sic a smack,
  They myed the joists a' bend an' crack.
  When deun, he took her roond the neck,
  An' faith, he didn't miss her, O.

The fiddler's elbow wagged a' neet,
He thought he wad dropped off his seat,
For deil a bit they'd let him eat,
They were sae keen on dancin, O.
  Some had to strip their coats for heat,
  An' shirts an' shifts were wet wi' sweat.
  They crammed their guts, for want o' meat,
  Wi' ginger-beer an' scranchim, O.

Now cocks had crawn an hour or more,
An' ower the yell-poy some did snore,
But how they luikt to hear the roar
O' Matt, the King Pit caller, O!
  'Smash him!' says Ned, 'he mun be wrang,
  He's callin through his sleep, aw's warr'n.'
  Then shootin to the door he ran;
  'Thou's asleep, thou rousty bawler, O!'

Now they danced agyen till it was day,
An' then went hyem – but, by the way,
Some o' them had rare fun, they say,
An' fand it nine months after, O.

Such tricks are played by heedless youth,
An' though they're common, north an' south,
That's nae excuse for breach o' truth,
Nor food for wit an' laughter, O.

Seun Wreckenton will bear the sway,
Two Members they'll put in, they say,
Then wor taxes will be deun away,
An' we'll a' sing now or never, O.
Baccy an' tea will be sae cheap,
Wives will sit up when they sud sleep,
An' we'll float in yell at wor pay week,
Then Wreckenton for ever, O!

yell = ale

Baggy Nanny, or the Pitman's Frolic

Come, lay up your lugs, and aw'll sing ye a sang,
It's nyen o' the best, but it's braw new and funny;
In these weary times, when we're not varry thrang,
A stave cheers wor hearts, though it brings us ne money.
Aw left Shiney Raw, for Newcassel did steer,
Wi' three or four mair of our neighbours se canny,
Determined te gan to the playhouse te hear
The king o' the fiddlers, the great Baggy Nanny.

We reached the Arcade, rather drouthy and sair –
It's a house full o' pastry-cooks, bankers and drapers;
At the fine fancy fare, hoo my marrows did stare
On the muffs, hats and beavers, se famed in the papers.
At Beasley's, where liquor's se cheap and se prime,
A bottle aw purchased for maw sweetheart Fanny,
We drank nowt but brandy – and when it wes time
We staggered away to see great Baggy Nanny.

We gat t' the door, 'mang the crowd we did crush,
Half-way up the stairs aw wes carried se handy –
The lassie ahint us cried, 'Push, hinny, push!'
Till they squeezed me as sma' and as smart as a dandy.
    We reached the stairheed, nearly smothered indeed;
    The gas-letters glistened, the paintin's looked canny;
    Aw clapped mesel doon 'side a lass o' reet breed,
    'Maw hinny,' says aw, 'hae ye seen Baggy Nanny?'

The lassie she twittered and looked rather queer,
And said, 'In this house there is mony a dozen.
They're planted se thick that there's ne sittin' here,
They smell se confounded o' catgut and rosin.'
    The curtain flew up, and a lady did squall,
    To fine music played by a Cockney bit manny.
    Then frae the front seats aw seun heard my friends bawl:
    'Hats off, smash yor brains, here comes greet Baggy Nanny!'

An ootlandish chep seun appeared on the stage,
And cut as odd capers as wor maister's flunkey.
He skipped and he fiddled as if in a rage –
If he had but a tail he'd a passed for a monkey.
    Deil smash a gud teun could this bowdykite play –
    His fiddle wad hardly e'en please my aud granny –
    So aw seun joined maw marrows, and toddled away,
    And wished a gud neet te the greet Baggy Nanny.

On crossin Tyne Brig, hoo wor lads ran the rig,
At bein se silly deun oot o' thor money.
Odd bother maw wig! had he played us a jig
We might telled them at hyem we'd seen somethin quite funny.
    But, law, be it spoke, and depend it's ne joke,
    Yen and a' did agree he was somethin uncanny –
    And dark ower each tree, before us he did flee,
    And fiddled us hyem, did the greet Baggy Nanny.

The Kippered Herrin'

'Boot pitmen an' keelmen thou's heard some queer jokes,
What wi' blunders, mistyeks, an' thor queer funny spokes,
For when we get a drop o' beer we're aal full o' glee,
Lads, we myek mony a blunder when we get on the spree.

Noo aw'll tell ye a trick we yence played on Jim Farrins,
Thit yen day bought a cask o' the best kippered herrins
To eat wiv his coffee, his taties an' breed,
Determined a' winter te hev a cheap feed.

He tuik fower greet big 'uns yen neet doon the pit,
An' he waddn't let us doon belaw taste a bit,
So a pennorth o' jalup we put iv his bottle,
An lads! hoo we laughed as it went doon his throttle.

He hewed half an hour till he felt summic ache,
Then he put doon his hands for te haud on the brake,
Cryin oot, 'Geordie Cairns, run away, thoo's maw cousin,
An' bring uz a doctor, for aw've swallowed some poison.'

Noo the bit trapper laddies, they laughed fit te borst,
An' mentioned what myed the poor man bad at forst,
But he says: 'Haud yor gobs an' give ower yor leein,
Aw's speechless aready, aw's sartin aw's deein.'

'Aw's deein, aw's deein, aw's off, Geordie Cairns,
Protect when aw's gyen maw poor wife an' bairns,
Keep a' maw pit claes, come draw thy lugs near,
An' hear maw last words, for aw've supped maw last beer.

‘Tell wor keeker aw deed wiv a pain i’ maw bowels,
Caused wi’ eatin some herrin aw bowt frae Jack Snowels;
Tell wor preacher next Sunday to pray for maw soul;
Tell wor owners an’ viewers aw’ll howk ne mair coal.

‘Tyek maw picks tiv aud Limpey an’ tell him aw’s gyen,
An’ come to maw funeral wi’ clogs ivery yen;
Tyek maw grandfether’s watch, keep that for theesel,
Aw’s gannin – ta ta, Geordie – to heaven or hell.’

The Cock-fight

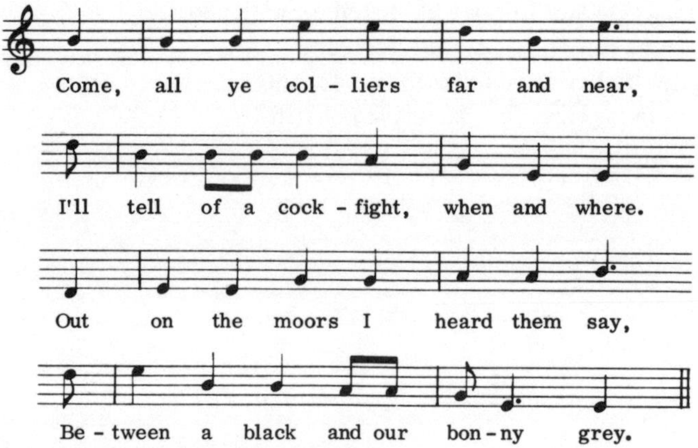

Come, all ye colliers far and near,
I'll tell of a cock-fight, when and where.
Out on the moors I heard them say,
Between a black and our bonny grey.

First come in was the Oldham lads;
They come with all the money they had.
The reason why they all did say:
'The black's too big for the bonny grey.'

It's into the pub to take a sup,
The cock-fight it was soon made up.
For twenty pound these cocks will play,
The charcoal-black and the bonny grey.

The Oldham lads stood shoutin round:
'I'll lay ye a quid to half a crown,
If our black cock he gets fair play,
He'll make mincemeat of the bonny grey!'

So the cocks they at it, and the grey was tossed,
And the Oldham lads said: 'Bah, you've lost!'
Us collier lads we went right pale,
And wished we'd fought for a barrel of ale.

And the cocks they at it, one, two, three,
And the charcoal-black got struck in the eye.
They picked him up, but he would not play,
And the cock-fight went to our bonny grey.

With the silver breast and the silver wing,
He's fit to fight in front of the king.
Hip hooray, hooray, hooray!
Away we carried our bonny grey.

Shovellin' Back the Slate

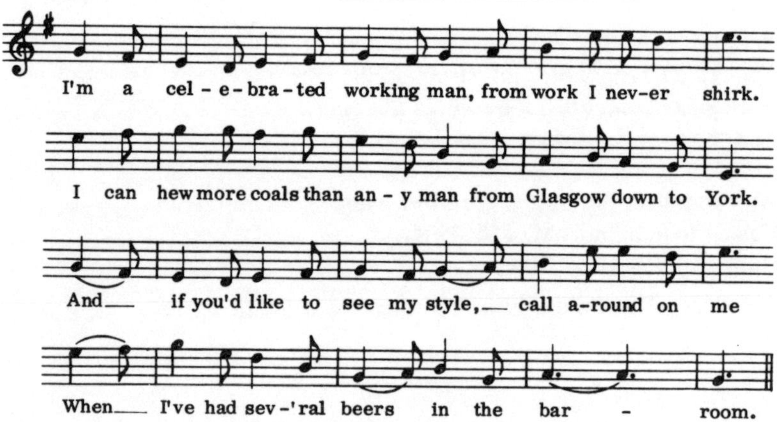

I'm a celebrated working man, from work I never shirk.
I can hew more coals than any man from Glasgow down to York.
And if you'd like to see my style, call around on me
When I've had several beers in the bar room.

> In the bar room, in the bar room, that's where we congregate,
> To drill the holes and fill the coals and shovel back the slate.
> And for to do a job of work, I am never late;
> That's provided that we do it in the bar room.

At puttin' I'm a dandy, I hope you will agree,
And gannin' along the gannin' board I mak the chum 'uns flee.
Your Kelly sweeps and back-ower torns they never bother me,
When I'm sittin' on the limmers in the bar room.

> In the bar room, in the bar room, that's where we congregate,
> To drill the holes and fill the coals and shovel back the slate.
> And for to do a job of work, I am never late;
> That's provided that we do it in the bar room.

I can judge a shot of powder to a sixteenth of a grain.
I can fill me eighteen tubs though the water pours like rain.
And if you'd like to see me in the porpendicular vein,
It's when I'm settin' timmer in the bar room.

> In the bar room, in the bar room, that's where we congregate,
> To drill the holes and fill the coals and shovel back the slate.
> And for to do a job of work, I am never late;
> That's provided that we do it in the bar room.

And now me song is ended, perhaps we'll have another.
Now don't you fire any shots in here or we will surely smother.
The landlord here would rather draw beer than go to all the bother
To put up the ventilators in the bar room.

> In the bar room, in the bar room, that's where we congregate,
> To drill the holes and fill the coals and shovel back the slate.
> And for to do a job of work, I am never late;
> That's provided that we do it in the bar room.

Stottin' doon the Waall

Today is Friday pay-day, an' the fyece will soon be off,
And Aa'm glad the week is ended, there's ne doot;
For we'll ride an' draw wor pay, an' then we'll de away,
Get wor dinner, put wor claes on, an' gan oot.

And it's stottin' doon the Waall,
Stottin' doon the Waall,
A Friday night is never quite complete
Unless Aa get me sup, an' Aa'm feelin' me way alang hyem
Be stottin' doon the Ropeworks Waall.

Noo, Aa gans doon te the corner, an' there Aa meets the lads,
An' we gan inte the Penny Wet for a sup;
An' we gets oot the bones, an' the tanners start te fly,
For when ye're suppin' ye never knaa yer luck.

Wey, later on again, when we're feelin' worsels lowsed,
We tyek worsels doon te the Mem.
As we prance aroond the floor, there's many a chance te score,
An' roond the neet off tyekin' a tart hyem.

Last Friday Aa did weel wiv little Nancy Broon,
Ye knaa her father works in 'Fowerth West';
As we're standin' in the door, Aa thowt te de a bit more,
But he oppened it up, an' ye'll guess where Aa went next.
Aye, Aa was stottin' alang the Waall,
Stottin' alang the Waall,
A Friday night is never quite complete
Unless Aa get me sup, an' Aa'm feelin' me way alang hyem
Be stottin' doon the Ropeworks Waall.

fyece will soon be off = work at the coal-face will soon be finished
ride = ride the conveyor belts
de away = go away
stottin' = bouncing, reeling
bones = dominoes
lowsed = finished, had enough
Mem = Wallsend Memorial Hall, where dances are held
Fowerth West = Fourth West, an area of the pit

III

Love and the Miner

III

LOVE AND THE MINER

When girls and men worked together in the dark lonely mines, opportunities for rough-and-ready courtship must have abounded. No songs survive to celebrate this. And though Frederick Engels declared that 'the number of illegitimate children in the coalfields is disproportionately large and indicates what goes on among the half-savage population underground', in fact seduction and its consequences, though traditionally a favourite theme of folk song, rarely appear in the balladry of the mines.

In British tradition, most love songs are for girls to sing. The songs may be full of joy or heartbreak, but the voice is usually that of a serious young woman. So it is with the songs of the coalfields. A girl sings her admiration for a youngster driving the coal waggons out of the pit and on to the loading staithes, and to her he is as handsome as any other, for all that he is pock-marked and one-eyed. Another girl echoes the old and universal folk song complaint of the bride ill-married, who wishes she were single again or married to a man in some other trade. *The Sandgate Girl's Lamentation* is one such, not concerned with miners but with keelmen who poled their laden barges from the staithes along the shallow Tyne and Wear to deeper water where the coal-boats lay. The keelmen were a wilder, more irresponsible race than the miners but they shared much of their life and many of their songs.

Where a ballad offers the picture of a broken-hearted bumpkin sighing for his girl, as in *Cushie Butterfield* (which has remained in favour for more than a century), it is usually a burlesque made for stage performance or to be accompanied by a clog-dance at some beery gathering. True folk song rarely treats love as farce. Among this class of songs, peculiar interest attaches to the broadside piece sometimes called *The Pitmen's Union* or *Brave Collier Lads*, in which trade union solidarity makes an early and rather shy appearance among the attributes desirable in a lover.

The Pitman's Union

It's on the fifth of April, the days were long and clear,
So bright did shine the sun and how cool did blow the air,
I overheard a pretty maid, how sweetly she did sing,
As she was sitting under her cow a-milking.

I stepped up to her and I bended my knee,
And asked her pardon for making so free.
'Your pardon is granted, young collier,' she replies,
'But do you belong to the brave union boys?'

'Don't you see that I'm a collier as black as any sloe?
And all this night I've been working down below.'
'Oh, I do love a collier as I do love my life,
For my father was a pitman all the days of his life.'

'Come all you noble gentlemen wherever that you be,
Oh, never pull their wages down to break their unity.
You see they hold like brothers, like sailors on the sea,
To do their best endeavours for their wives and family.'

'We've steam upon the ocean, we've steam upon the land,
And what can we do now at all without our collier men?
They send the coal above the ground, the country all around;
We know they work both day and night with danger all around.'

She put her arms around him like violets round the vine.
'You are the bonny collier lad that's won this heart of mine.
And since that you have won the day and since you've won my heart,
I'll crown you in glory and for ever take your part.'

Come all you pretty fair maids, wherever that you be,
Never despise a collier lad in any one degree.
For if that you do use them well, they'll do the same to thee.
There's no one in this world but a bonny pit lad for me.

Brave Collier Lads

As I walked forth one summer's morn, all in the month of June,
The flowers they were springing, and the birds were in full tune,
I overheard a lovely maid, and this was all her theme:
'Success attend the collier lads, for they are lads of fame.'

I stepped up to her, and bending on my knee,
I asked her pardon for making with her so free.
'My pardon is granted, young collier,' she replies,
'Pray do you belong to the brave Union boys?'

'You may see I'm a collier as black as a sloe;
And all the night long I am working down below.'
'Oh I do love a collier as I do love my life –
My father was a pitman all the days of his life.

Come now, my young collier, and rest here awhile,
And when I have done milking I'll give you a smile.'
He kissed her sweet lips while milking her cow;
And the lambs they were sporting all in the morning dew.

Come all you noble gentlemen, wherever you may be,
Do not pull down their wages, nor break their unity;
You see they hold like brothers, like sailors on the sea,
They do their best endeavours for their wives and family.

Then she clapt her arms around him like Venus round the vine.
'You are my jolly collier lad: you've won this heart of mine;
And if that you do win the day, as you have won my heart,
I'll crown you with honour, and for ever take your part.'

The colliers are the best of boys, their work lies under ground,
And when they to the alehouse go they value not a crown;
They spend their money freely and pay before they go;
They work under ground while the stormy winds do blow.

So come all you pretty maidens wherever you may be,
A collier lad do not despise in any degree,
For if that you do use them well they'll do the same to thee;
There is none in this world like a pit-boy for me.

A Pitman's Love Song

Aw wish my lover she was a cherry,
Growing upon yon cherry tree,
And aw mysel a bonny blackbird;
How aw would peck that cherry cherree.

Aw wish my lover she was a red rose,
Growing upon yon garden wa',
And aw mysel was a butterflee;
Oh, on that red rosie aw wad fa'.

Aw wish my lover she was a fish,
Sooming doon in the saut sea,
And aw mysel was a fisher lad;
Oh, aw wad catch her reet cunningly.

Aw wish my love was in a kist,
And aw mysel to carry the key.
Then aw wad gan tiv her when aw had list,
And bear my hinny good company.

Aw wish my love she was a grey ewe,
Grazing by yonder river side,
And aw mysel a bonny black tup;
Oh, on that ewie how aw wad ride.

Oh, she's bonny, she's wondrous canny.
Oh, she's well far'd for to see.
For the mair I think on my love's on upon her,*
And under her apron fain wad aw be.

* Another song has: 'The more I think on her, my love's strong upon her'. Perhaps something of the sort is meant here.

Aw wish my lover she was a bee skep,
And aw mysel a bumble bee,
That aw might be a lodger within her;
She's sweeter than honey or honeycomb tea.

Aw wish my lover was a ripe turd,
Smoking doon in yon dyke side,
And aw mysel was a shitten flee;
Aw'd sook her all up before she was dried.

O my hinny, my bonny hinny,
O my hinny, my bonny hinnee;
The mair aw think on her, my heart's set upon her.
She's fairer than ever she used to be.

The Bonny Pit Laddie (I)

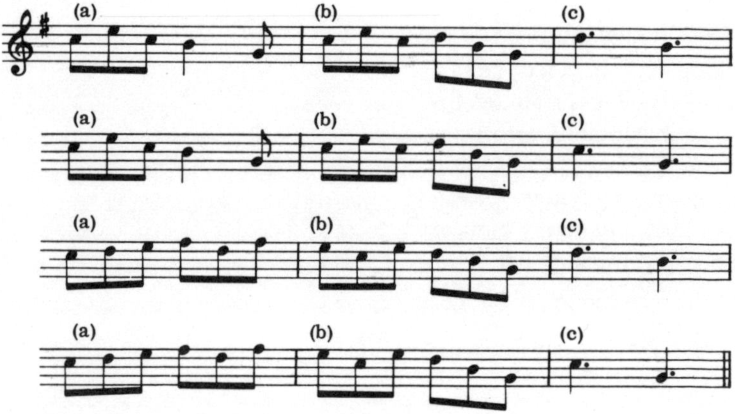

The bonny pit laddie, the canny pit laddie,
  The bonny pit laddie for me, O.
He sits in a hole as black as the coal
  And brings the white siller to me, O.

The bonny pit laddie, the canny pit laddie,
  The bonny pit laddie for me, O.
He sits on his cracket and hews in his jacket,
  And brings the bright siller to me, O.

The Bonny Pit Laddie (II)

The bonny pit laddie, the canny pit laddie,
The bonny pit laddie for me, o.
He sits on his cracket as black as a jacket
An' brings the siller to me, o.
    He works hard, he brings me all his money, o.
    He taks me in his arms an' he calls me his hinny, o.

The bonny pit laddie, the canny pit laddie,
The bonny pit laddie for me, o.
He sits on his hunkers and yacks at the bunkers,
An' just brings the coppers to me, o.
    Now, he comes hyem an' he tells me all his troubles, o.
    He's had a row wi' the gaffer about his lazy marras, o.

The bonny pit laddie, the canny pit laddie,
The bonnie pit laddie for me, o.
Now he's had his yokins an' keeps all his tokens,
An' brings the white siller to me, o.
Now he's all reet with all the men and gaffers, o.
An' he comes hyem an' he calls me his hinny, o.

The Collier Lad

A collier lad, a collier lad,
   A collier lad for me, O.
He works in a hole as black as the coal,
   And brings all his brass home to me, O.

He says: 'Sithee, lass, Ah've browt thi me brass,
   Ah'll niver get drunk ony more, O.
Ah'll save all me brass and buy a jackass,
   And ride past the ale-house door, O.'

Collier Lads

My mother sent me for some water
For some water for my tea.
My foot slipped and down I tumbled.
Along came a collier lad a-courtin me.

Collier lads gets gold and silver.
Factory lads gets nobbut brass.
Who'd get married to a two-loom weaver
When there's plenty o' collier lads.

A collier lad, a collier lad,
A collier lad for me.
He works in a hole as black as the coal
And he brings a' his money to me.

I'll Have a Collier for my Sweetheart

My mother said I must not have a collier,
For if I do he'll surely break my heart.
But I don't care what my mother tell me,
I'll have a collier for my sweetheart.

Collier lads get gold and silver,
Factory lads get nowt but brass.
Who'd be bothered with a naughty bobber,
When there's plenty collier lads?

If you'll leave your collier sweetheart,
I'll buy thee a guinea gold ring.
I'll buy thee a silver cradle,
For to rock thy baby in.

I don't want your silks and satin,
Nor your guinea golden ring.
I don't want a silver cradle
For to rock my baby in.

My mother said that I could be a lady,
If from my collier lad I'd quickly part;
I'd sooner walk the bottom of the ocean
Than I'd give up my collier sweetheart.

The Waggoner

Saw ye owt o' ma lad,
Gannin' doon the waggon way,
Wiv his pocket full o' money
And his bag full o' hay?
    Oh, ma lad's a canny lad,
    The canniest Aa see,
    Though he's sair frowsy-freckled
    And he's blind of an e'e.

There's nivor a lad like ma lad,
Drives to the staithes on Tyne.
He's coaly black on work-days
But on holidays he's fine.
    Oh, ma lad's a canny lad,
    The canniest Aa see,
    Though he's sair frowsy-freckled
    And he's blind of an e'e.

Wiv his siller in his hand,
Wiv his love in his e'e,
Yonder a see ma canny lad
A-comin' doon to me.
    Oh, ma lad's a canny lad,
    The canniest Aa see,
    Though he's sair frowsy-freckled
    And he's blind of an e'e.

The Sandgate Girl's Lamentation

I was a young maid truly,
And lived in Sandgate Street;
I thought to marry a good man
To keep me warm at neet,
    Some good-like body, some bonny body
    To be with me at noon;
    But last I married a keelman,
    And my good days are done.

I thought to marry a parson
To hear me say my prayers,
But I have married a keelman
And he kicks me down the stairs.
    He's an ugly body, a bubbly body,
    An ill-far'd ugly loon;
    And I have married a keelman
    And my good days are done.

I thought to marry a dyer
To dye my apron blue,
But I have married a keelman
And he makes me sorely rue.
    He's an ugly body, a bubbly body,
    An ill-far'd ugly loon;
    And I have married a keelman
    And my good days are done.

I thought to marry a joiner
To make me chair and stool,
But I have married a keelman
And he's a perfect fool.
    He's an ugly body, a bubbly body,
    An ill far'd ugly loon;
    And I have married a keelman
    And my good days are done.

I thought to marry a sailor
To bring me sugar and tea,
But I have married a keelman
And that he lets me see.
    He's an ugly body, a bubbly body,
    An ill-far'd ugly loon;
    And I have married a keelman
    And my good days are done.

The Collier Laddie

Oh, I've been east and I've been west,
And I've been to Kirkcaldy,
But the bonniest lass that ever I saw
She was followin a collier laddie.

She'd silken slippers on her feet,
Her body neat and handsome;
She'd a sky-blue ribbon in her hair,
And her eyes like gowd were glancin.

'Oh, where live ye, my bonny lass,
And tell me what they ca' ye?'
'Bonny Jean Gordon is my name,
And I'm followin my collier laddie.'

'Oh, see ye not yon high, high hills,
The sun shines on sae brawly?
They are a' mine, and they'll be thine,
Gin ye leave your collier laddie.

And ye will busk in gay attire,
   Weel busket up sae gaudy,
And ane to wait on every hand
   Gin ye leave your collier laddie.'

'Though ye had a' the sun shines on,
   And the earth conceals sae lowly,
I would turn my back on you and it a',
   And embrace my collier laddie.'

But he has to her father gane,
   To her father gane sae brawly.
'Ye'll mak your daughter wed wi' me,
   And forsake your collier laddie.

I'll gie her lands, I'll gie her rents,
   And I'll mak her a lady.
I'll mak her ane o' a higher degree
   Than to follow a collier laddie.'

Then he has to his daughter gane,
   To his daughter gane sae brawly,
Says: 'Ye'll gang wi' this braw gentleman,
   And forsake her collier laddie.

He'll gie ye lands, he'll gie ye rents,
   And he'll mak ye a lady.
He'll mak ye ane o' a higher degree,
   Than to follow a collier laddie.'

'I winna hae his lands an' I winna hae his rents,
   And I winna be his lady,
I'll mak my bed in the collier's neuk,
   An' lie doon wi' my collier laddie.'

When twenty years hae come an' gane,
   Twenty years sae brawly,
The same silly gentleman cam' beggin his breid,
   Frae Jean an' her collier laddie.

'Whaur's a' the lands, an' whaur's a' the rents,
   That was tae make me a lady?
I've got gold and gear eneuch
   An' I'm aye wi' my collier laddie.'

Noo she is tae her father gane,
   Tae her father gane sae brawly,
Says: 'What ye's to gie to your braw gentleman,
   Ye'll gie to my collier laddie.'

He's counted doon five hundred pounds,
   Five hundred pounds sae brawly,
Says: 'Tak ye that, my daughter Jean,
   An' gie't tae your collier laddie.'

'Love for love is a bargain for me,
   Though the wee cot-hoose should haud me,
An' the world before tae win my breid,
   An' share wi' my collier laddie.'

Jimmy's Enlisted
(or: The Recruited Collier)

Oh, what's the matter wi' you, my lass,
An' where's your dashin Jimmy?
The sowdger boys have picked him up
An' sent him far, far frae me.
      Last pay-day he set off to town,
      An' them red-coated fellows
      Enticed him in an' made him drunk,
      An' he'd better gone to the gallows.

The very sight o' his cockade,
It set us all a-cryin;
An' me, I fairly fainted twice,
I thought that I was dyin.

My father would ha' paid the smart,
An' he run for the golden guinea,
But the sergeant swore he'd kissed the book,
And now they've got young Jimmy.

When Jimmy talks about the wars,
It's worse than death to hear him.
I must go out an' hide my tears,
Because I cannot bear him.
For aye he jibes an' cracks his jokes,
An' bids me not forsake him.
A brigadier or grenadier
He says they're sure to make him.

As I walked o'er the stubble field,
Below it runs the seam,
I thought o' Jimmy hewin there,
But it was all a dream.
He hewed the very coals we burn,
An' when the fire I's leetin,
To think the lumps was in his hands,
It sets my heart to beatin.

So break my heart, and then it's ower,
So break my heart, my dearie,
An' I'll lie in the cold, cold grave,
For of single life I'm weary.

Jolly Joe the Collier's Son

I'm Jolly Joe the collier's son, near Oldbury town I dwell.
I courted lasses many one, and I loved them all right well.
I courted Nancy and young Kate and buxom young Nelly too
But Rachel is the girl I adore and that you soon shall know.

Come all your colliers in this row, who delight in a bonny lass,
Who loves to drink ale that's brown that sparkles in the glass,
My parents they do frown on me and say I am to blame
For keeping Rachel's company who liveth in Marsh lane.

When I rose up one morning at the dawning of the day
I like to hear the small birds sing, see the lambs to skip and play.
I took a walk to Oldbury town round by the Bilston hill
And there I spied my own true love with Jack of Armlow mill.

I hid myself behind a shade a distance from whence they were.
He gave her kisses one, two, three, not knowing I was there.
I boldly stepped up to them saying: Rogue, what hast thou done?
I'm Jolly Joe the collier's son so you must either fight or run.

Hold your hand, dear Joe, she said, and no more of that let's have,
I'll be thy servant, slave and wife till we both go to one grave.
Then to the church young Rachel went right sore against her will
So maidens all, pity my downfall by Jack from Armlow mill.

The Sandgate Lass on the Ropery Banks

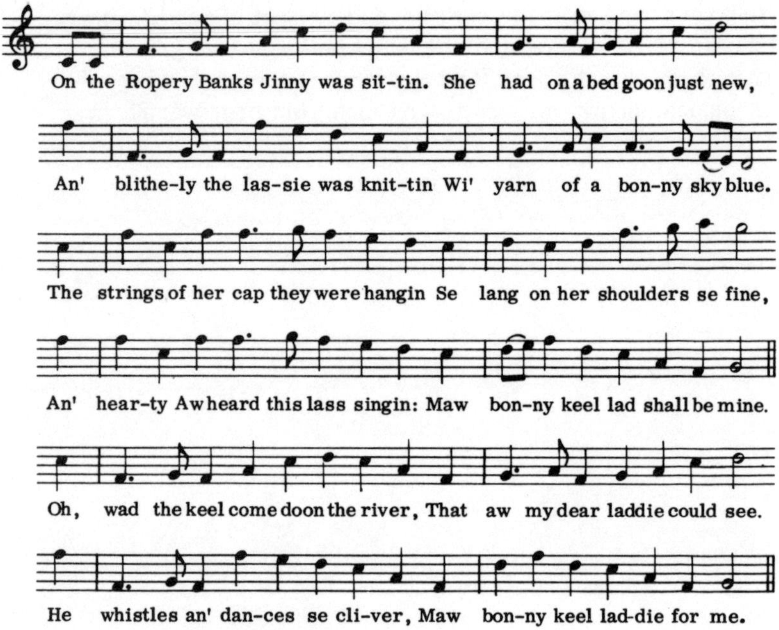

On the Ropery Banks Jinny was sittin.
  She had on a bed-goon just new,
An blithely the lassie was knittin
  Wi' yarn of a bonny sky blue.
The strings of her cap they were hangin
  Se lang on her shoulders se fine,
An' hearty Aw heard this lass singin:
  Maw bonny keel lad shall be mine.
    Oh, wad the keel come doon the river,
    That Aw my dear laddie could see;
    He whistles and dances se cliver,
    Maw bonny keel laddie for me.

Last neet in among these green dockins
  He fed me wi' gingerbread spice.
Aw promised te knit him these stockins.
  He cuddled an' kissed me se nice.
He ca'ed me his jewel an' his hinny.
  He ca'ed me his pet an' his bride,
An' he swore that Aw should be his Jinny,
  To lie at neets doon by his side.

That mornin forget Aw will niver,
  When first Aw saw him on the Quay,
The *Keel Row* he whistled se cliver,
  He won my affections fre me.
His drawers on his doup luiked se canny,
  His keel hat wes cocked on his heed,
An' if Aw'd not getten maw Jimmy,
  Faith, by this time Aw wad been deed.

The first time Aw spoke te maw Jimmy
  – Now mind ye, it isn't a lee –
Maw mother had gi'en me a penny
  Te bring her a penn'orth o' tea,
When a lad i' the street cried oot: 'Bessie!'
  Says I: 'Hinny, that's not maw nyem.'
'Becrike, niver mind,' he says. 'Lassie,
  Toneet Aw will see ye syef hame.'

Since then Aw have been his true lover,
  Aw've loved him as dear as maw life,
An' in spite o' byeth fether an' mother,
  Aw'll suen be maw keel-laddie's wife.
Hoo happy we'll be then tegither,
  When he brings hyem his wages te me,
Wiv his bonny bit bairn cryin: 'Fether,'
  An' another one laid o' maw knee.
    Oh, wad the keel come doon the river,
    That aw maw dear laddie could see;
    He whistles an' dances se cliver,
    Maw bonny keel laddie for me.

Sandgate Wife's Nurse Song

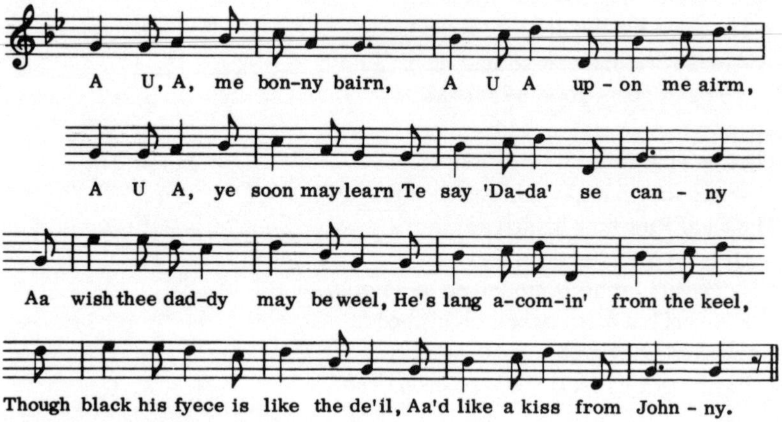

A U A, me bonny bairn,
A U A, upon me airm,
A U A, ye soon may learn
Te say 'Dada' se canny.
    Aa wish thee daddy may be weel;
    He's lang a-comin' from the keel.
    Though black his fyece is like the de'il,
    Aa'd like a kiss from Johnny.

Thoo really has thee daddy's chin.
Thoo'rt like him, baith leg and limb.
Aa wi' pleasure will thee sing,
Thoo's a'together bonny.
    Noo, Johnny he's a clever lad –
    Last night he fuddled a' he had;
    The morn he wasn't very bad,
    He looked as blithe as ony.

Though thoo's the forst, thoo's not the last.
Aa mean te hae me bairnies fast.
When this happy time is past,
Aa still will love me Johnny.
    For his hair is brown and so is thine,
    Thee eyes are grey and so are mine;
    Thee nose is tapered off se fine –
    Thoo's like thee daddy, Johnny.

Thee canny doup is fat and round,
And like thee dad, thou's plump and sound.
Thou's worth te me ten thoosand pound;
Thoo's altegither bonny.
    When daddy's drunk, he'll tak a knife,
    Threaten sore te tak me life;
    But who wadden't be a keelman's wife
    Te have a lad like Johnny.

But yonder daddy's comin' noo.
He's the best thing amang the crew.
He's gone te the Barley Mow,
Me canny god-like Johnny.
    Come, let's get the bacon fried.
    Let us make a clean fireside.
    Then on his knee thoo's get a ride
    When he comes hyem te mammy.

A U A, me bonny bairn,
A U A, upon me airm,
A U A, ye soon may learn
Te say 'Dada' se canny.

The Rambling Miner

I am a miner stout and bold,
Long time I've worked down underground,
To raise both tin and copper, too,
For the honour of our miners.
    Now brother miners, I'll bid you adieu,
    I'll go to work no more with you,
    But scour the country through and through,
    And still be a rambling miner.

Now when I came to Redruth Town,
There I saw lasses plenty;
I boldly stepped right up to one,
To court her for her beauty.
    I said, 'My dear, be of good cheer;
    I'll never leave you, have no fear;
    But travel the country through and through,
    And still be a rambling miner.'

And when I come to Mertha Town,
There I saw lasses bonny;
I boldly stepped right up to one,
To court her for her money.

I said, 'My dear, be of good cheer;
There's ale and wine and rum punch, too,
Besides a pair of new silk shoes,
To travel with a rambling miner.'

It was in the morning when I rose,
I left my love a-sleeping;
I left her for an hour or two,
While I went to court some other.
If she stays there till I return,
She may stay there till the day of doom;
I'll court some other girl in her room,
And still be a rambling miner.

The very first house I went up to,
They made me lie out in the barn;
A pretty young girl to her mother did say,
'I'm afraid you will do him some great harm
If you put him to lie on the cold barn floor,
Where the wind and rain comes in the door;
You may pop him in my bed, mother, if you please,
For I'm told he's a rambling miner.'

Now when I came to Callington,
I met with Captain Thomas;
I asked him for a job of work,
Which he gave me the promise.
He told me I should have a job,
To work in the shaft with pick and gad;
And good wages he'd allow to me,
For being a rambling miner.

And if you want to know my name,
My name it is young Simmons;
I've a roving commission from the queen
To court all girls who're handsome.
With my false heart and deluding tongue
I'll court them all, both old and young;
I'll court them all and marry none,
And still be a rambling miner.

Cushie Butterfield

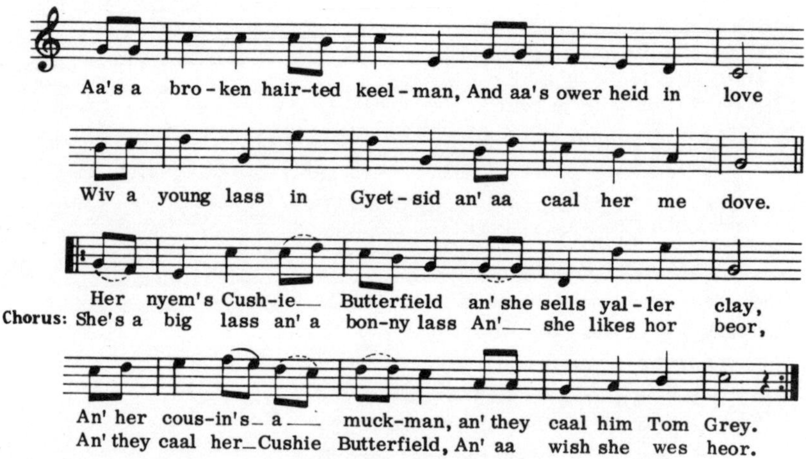

Aa's a broken-hairted keelman, and Aa's ower heid in love
Wiv a young lass in Gyetside an' aa caal hor me dove.
Hor nyem's Cushie Butterfield, an' she sells yaller clay,
An' hor cousin is a muckman, an' they caal him Tom Grey.
  She's a big lass, an' a bonny lass,
    An' she likes hor beor,
  An' they caal hor Cushie Butterfield,
    An' Aa wish she wes heor.

Hor eyes is like two holes in a blanket bornt through,
Hor broos in a mornin wad spyen a young coo,
An' when aa hear hor shoutin: 'Will ye buy ony clay?'
Like a candyman's trumpet, it steals me heart away.

Ye'll oft see her doon at Sangit when the fresh harrin comes in.
She's like a bag o' sawdust, tied roond wiv a string.
She wears big galoshes tee, an' hor stockins once wes white,
An' hor bedgoon it's laelock, an' hor hat's nivor strite.

When Aa axed hor te marry me she started te laugh;
'Noo, nyen o'yer monkey tricks, for aa like ne sich chaff.'
Then she started a-blubbin, an' she roared like a bull,
An' the cheps on the Quay says Aa's nowt but a fyeul.

She says the chep that gets hor must wark ivory day,
An' when he comes hyem at neets he must gan an' seek clay;
An' when he's away seekin, she'll myek baals an' sing:
O weel may the keel row that ma laddie's in.
    She's a big lass, an' a bonny lass,
        An' she likes hor beor,
    An' they caal hor Cushie Butterfield,
        An' Aa wish she wes heor.

yaller clay = a mixture of yellow clay and pumice, rolled into balls and sold for cleaning and brightening doorsteps

Six Jolly Miners (I)

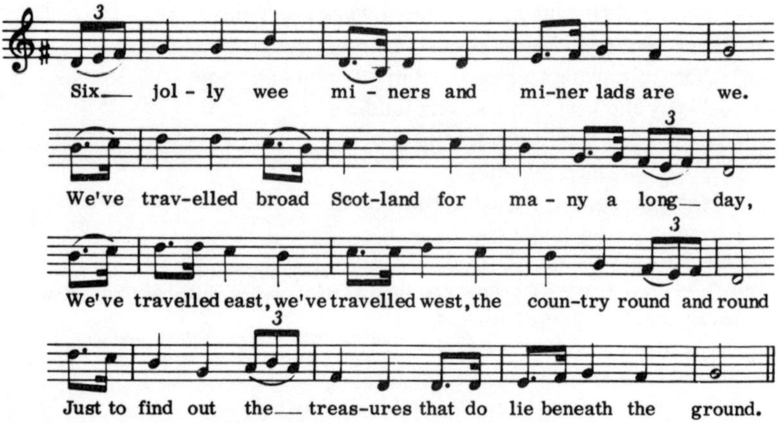

Six jolly wee miners, and miners lads are we,
We've travelled broad Scotland for many a long day.
We've travelled east, we've travelled west, the country round and round
For to find out the treasures that lie below the ground.

You should see my miner lad as he walks down the street,
All dressed in his best, lookin gentle and sae neat.
His teeth are white as ivory, his eyes are black as sloes.
You can easy tell a miner lad everywhere he goes.

Some hae got money, and some hae none at a',
And them that hae money, the bottle they will draw.
We'll fill our glasses to the tip, the toast goes round and round,
My health to every wee collier lad that works below the ground.

I'll knit my love a grovet, as doggie as it can be,
The colours that I'll put in it will fairly tak his e'e.
His chums will a' come up to him, and say: 'Where'd ye get that?'
'Oh, I got it frae ma wee doggie moll, an' whit dae ye think o' that?'

I'll build my love a castle, a castle of renown,
Neither kings, dukes nor earls will pull my castle down.
The king loves the queen an' the emperor does the same,
I love my wee miner lad an' wha can me blame?

Six Jolly Miners (II)

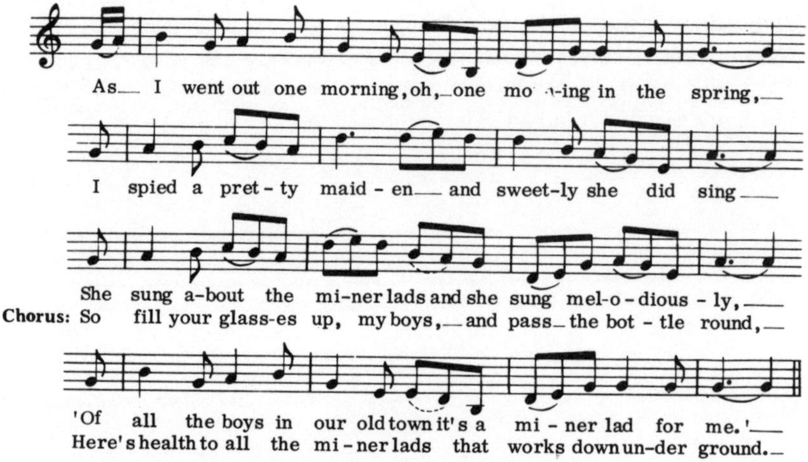

As I went out one morning, oh, one morning in the spring,
I spied a pretty maiden and sweetly she did sing.
She sung about the miner lads and she sung melodiously,
'Of all the boys in our old town, it's a miner lad for me.'
> So fill your glasses up, my boys, and pass the bottle round.
> Here's a health to all the miner lads that works down underground.

There's two of 'em came from Ballycastle and two from Benbrack town,
Two more of 'em came from Killarney, and that's a place of great renown.
They're six as fine a fellows as ever you did behold,
And they can turn the hardest rock into sovereigns of bright gold.

Oh, the farmer takes delight in the sowing of his corn.
The huntsman takes delight in the blowing of his horn.
The miner takes delight for to strike the rock on down
And to find out the treasure that lies below the ground.

Oh, it's teeth as white as ivory and eyes as black as sloes.
It's easy to know a miner lad wherever that he goes.
And sometimes they've got money, but more times they have none,
But here's a health to all them miner lads that works down
underground.

My Collier Lover

Once I had a collier lover,
And he thought the world of me,
Till one day he met another.
Now he thinks no more of me.

Take this ring from off my finger,
Take this necklace off my neck,
Take it to that other lover,
To the one you love the best.

Brothers, they have all forsook me,
Sisters, they have done the same.
Now there's only Dad who loves me;
Mother hangs her head in shame.

When I'm on my bed of sickness,
When I'm on my bed of shame,
He will come and ask forgiveness.
He will want me back again.

Goodbye Dad, and goodbye Mother,
Goodbye all my mining friends,
Goodbye you, my collier lover,
You have brought me to this end.

If by any chance you'll meet him
Always greet him with a smile,
Because you know I loved him dearly,
He's the father of my child.

Ma Lad's a Canny Lad

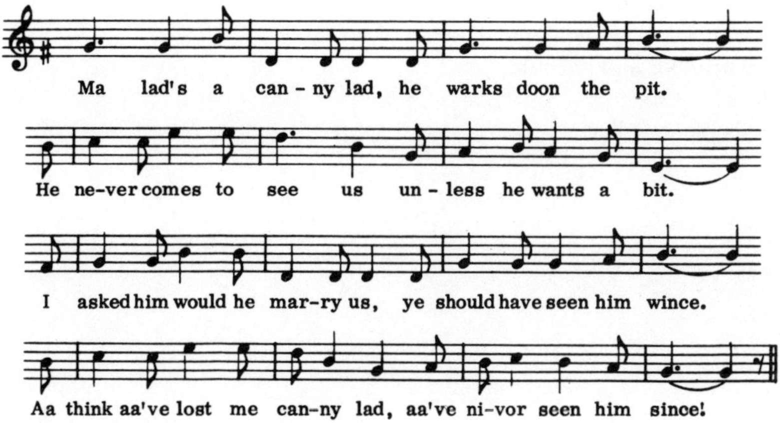

Ma lad's a canny lad, he warks doon the pit.
He never comes to see us unless he wants a bit.
I asked him would he marry us, ye should have seen him wince.
Aa think aa've lost me canny lad, aa've nivor seen him since!

Pick, Pick, Pick

Pick, pick, pick,
That's the music down the mine.
I learned to play the tune
When I was scarcely nine.
Today it is my wedding day
So let the bottle pass,
And drink good health and plenty of wealth
To a collier and his lass!

IV

Boy Miners and Old Miners

IV

BOY MINERS AND OLD MINERS

There are miners not long retired who started in the pits at the age of eight, and little over a century ago small children were to be seen dragging tubs along the ways. The smallest boys of all – the six- and five-year-olds – sat in the dark passages ready to open and shut the trap-doors regulating the ventilation of the mine, to allow coal and colliers to pass through. Sometimes on their way home from the pit the weary children would lie down, to be found by their parents late at night, asleep by the roadside.

The pathos of the trapper-boys who 'sit and keep a door midst the darkness and the gloom', and of the white-haired miners turned away from the pit and thrust into what was ironically called 'the colliers' second childhood', picking slate on the bank alongside little boys, was something that several stage comedians exploited in their sentimental songs. The miner himself knew better than any professional song-writer the true meaning of the old pitman's plight when, after a lifetime of hard work underground, he found himself facing poverty and the poor-house. So the miners' own treatment of the theme was plain and factual and not without dignity, as in *The Banks of the Dee*, much sung even today.

Fourpence a Day

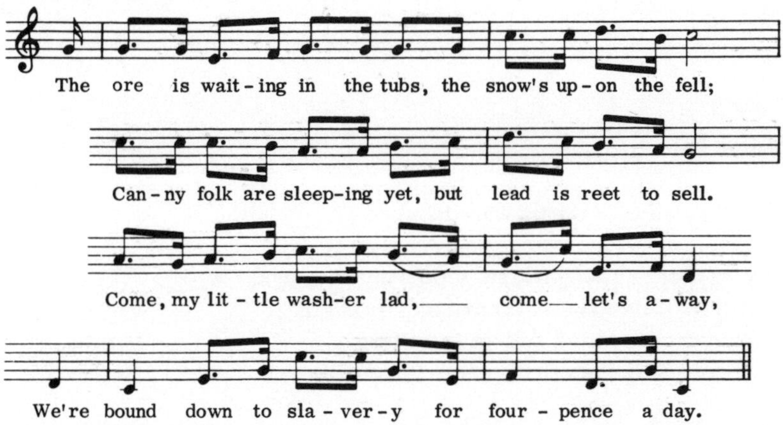

The ore is waiting in the tubs, the snow's upon the fell;
Canny folk are sleeping yet, but lead is reet to sell.
Come, my little washer lad, come let's away,
We're bound down to slavery for fourpence a day.

It's early in the morning we rise at five o'clock,
And the little slaves come to the door to knock, knock, knock.
Come, my little washer lad, come let's away,
It's very hard to work for fourpence a day.

My father was a miner and lived down in the town,
'Twas hard work and poverty that always kept him down.
He aimed for me to go to school, but brass he couldn't pay,
So I had to go to the washing rake for fourpence a day.

My mother rises out of bed with tears on her cheeks,
Puts my wallet on my shoulders which has to serve a week.
It often fills her great big heart when she unto me does say:
'I never thought thou would have worked for fourpence a day.'

Fourpence a day, my lad, and very hard to work,
And never a pleasant look from a gruffy-looking Turk.
His conscience may it fail and his heart may it give way,
Then he'll raise us our wages to ninepence a day.

Little Pal

Little pal, thoo's startin wark teday.
There's one or two things that Aa want te say:
De as thee mother says, an' thoo'll aalways win,
But it aal depends on the cavil that thoo's in.
Aa knaa thoo hasn't had an easy start,
So here's a thought that's aalways in my heart:

> Aa want tha te hew, little pal,
> Te get one or two, little pal,
> Aa want tha te put, te corve and te drill,
> Te take up some canch so tha father can fill.
> An' when thoo gans for thee pay, little pal,
> Aa hope things'll be O.K., little pal.
> If thoo draws three pund ten,
> That's two quid for the aad hen,
> Thorty bob for theesen, little pal.

Geordie Black

Ma name is Geordie Black, aa'm gettin' very aad.
Aa've hewed tons of coal in me time,
An' when aa wes just a lad, aa could either put or hew.
Oot the others aa could aalways tak the shine.
Aa'm gannin doon the hill, I cannet use the pick.
The maister hes ne pity on aad bones.
So noo aa'm on the bank, an' aa while me time away
Amang the bits o' lads wi' pickin' oot the stones.
    Oh, me name is Geordie Black. In me time aa've been a crack.
    An' aa've warked baith in the Gyuss an' in the Betty.
    An' the coals upon the Tyne oot the others tak the shine,
    An' we lick 'em aal for iron doon at Hawks's.

Noo, when aa wes just a lad carried on me father's back,
An' away he wad take me te the pit,
An' to get in the cage an' te gan doon belaw
'Twas enough te mak a youngster tak a fit.
For te sit an' keep a door midst the darkness an' the gloom,
Ay, mony an 'oor be messel,
An' te hear the aaful shots that rummelled roond the pit,
An' lumps o' roondy coal come doon pellmell.
    Oh, me name is Geordie Black. In me time aa've been a crack.
    An' aa've warked baith in the Gyuss an' in the Betty.
    An' the coals upon the Tyne oot o' the others tak the shine,
    An' we lick 'em aal for iron doon at Hawks's.

Noo aa'll say good neet, for it's nearly time te lowse.
An' aa've done me best te please ye, ivery one.
Young lads that's here the neet, mind, de the thing that's reet.
In this warld that's the way te get on.
So here's successs te trade baith on the Wear an' Tyne!
Aa divent like te see the place se slack.
For if wor pit lies idle, ne pay-note comes teneet,
An' it grieves the heart o' poor Geordie Black.
    Oh, me name is Geordie Black. In me time aa've been a crack.
    An' aa've warked baith in the Gyuss an' in the Betty.
    An' the coals upon the Tyne oot o' the others tak the shine,
    An' we lick 'em aal for iron doon at Hawks's.

An Old Miner

Oh, once I was a collier with pride in me work.
I worked the face at Moston; from hard graft I never shirked.
But they closed the pit around us and sent us all away.
I was too old for other work, so I walk the roads today.

It's hard when once you've had some pride in how you earned your brass,
An' never had to look to none, an' never had to ask.
For now they've shut us out of t' pit I've 'ad to beg an' scrape.
The things I feel, the things I've done, I hardly like to say.

I smashed me back when t' roof fell in. They never thowt I'd lake,
But with hungry children wantin' bread, I had to for their sake.
I lost an eye to flyin slag, me lungs are full of dust,
Me body's scarred, me bones are bent, but me heart it feels it worst.

You'd think a man who gave his life to workin' down below
Would 'ave a better thanks than this, and not be brought so low.
I don't ask more than is me due, than anyone would expect.
Just give me back what I deserve – a livin', an' respect.

'You workin' men's much better off', that's how they tell you lies.
But what's the use of livin' when a man has lost his pride?
Don't treat us like machines and turn us off when you have done.
It's only pride in honest work that lets us carry on.

Me name it is John Connelly, at t' pit I knew me place.
But they took me labour off me an' they threw it in me face.
I dunnot ask for ower much, it's just that you should think
How much his work means to a man, before you let him sink.

lake = live

The Banks of the Dee

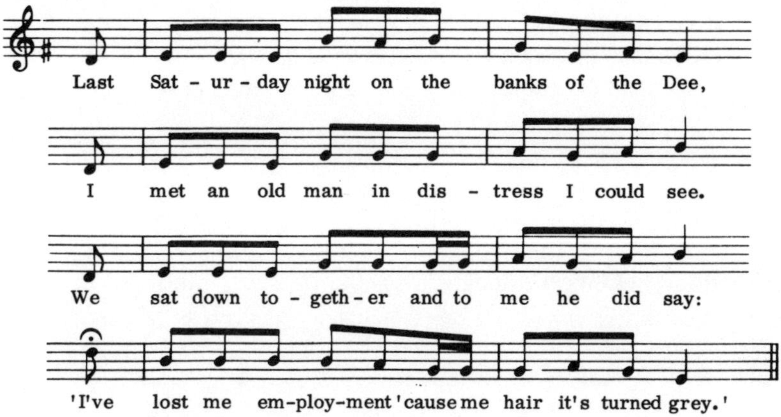

Last Saturday night on the banks of the Dee,
I met an old man, in distress I could see.
We sat down together and to me he did say,
I've lost my employment 'cause my hair it's turned grey.
    I am an old miner, aged fifty and six,
    If I could get lots, I would raffle my picks,
    I'd raffle them, sell them, I'd throw them away,
    For I can't get employment, my hair it's turned grey.

When I was a young chap I was just like the rest,
Each day in the pit I'd do my very best.
When I had a loose place, I'd be filling all day,
Now at fifty and six, my hair it's turned grey.

Last Wednesday night to the reckoning I went.
To the colliery office, I went straight fornenst.
I'd just got my pay packet, I was walking away,
When they gave me my notice, 'cause my hair it's turned grey.

Now all you young fellows, it's you that's to blame.
If you get good places, you'll do just the same.
If you get good prices, you'll hew them away,
But you're sure to regret it when your hair it's turned grey.
    I am an old miner, aged fifty and six,
    If I could get lots I would raffle my picks,
    I'd raffle them, sell them, I'd throw them away,
    For I can't get employment, my hair it's turned grey.

The Old Clog-Dancer's Song

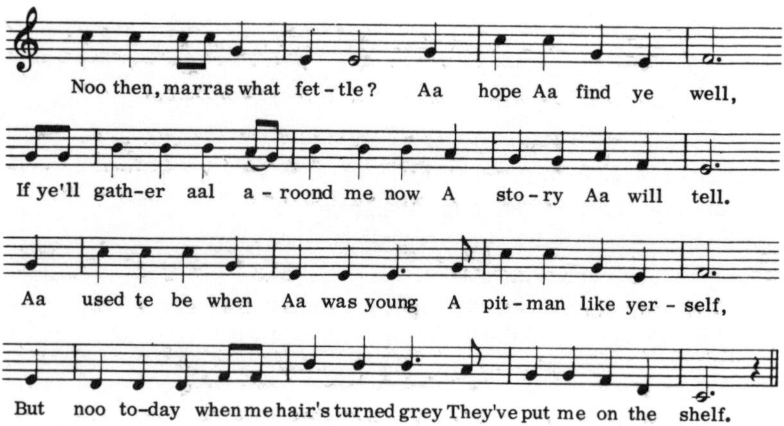

Noo then, marras, what fettle?
Aa hope Aa find ye well.
If ye'll gather aal aroond me now
A story Aa will tell.
Aa used te be when Aa wes young
A pitman like yerself,
But noo today when me hair's turned grey
They've put me on the shelf.
    For fifty years Aa've worked in the pits
    And mony's the change Aa've seen.
    When Aa wes young and a man wes done

They set him on the screens.
But noo today, when yer hair turns grey,
Like a squaddie with his aad peg leg,
They'll tell ye ye're ne use ony mair,
Get oot on the road and beg.

Noo, me age it is but sixty-two,
And for work Aa've got ne chance.
They've barred me from the championships,
Wi' me clogs for the belt te dance.
They said that Aa wes gettin' aad
And for dancin' Aa wes ne use.
They telled me te hoy me clogs away,
Get into the aad workhoose.

So come aal ye young fellers,
Take heed what Aa say:
Get oot o' the pits while ye're young enough
To make good your own way
For if ye divvent and ye stay belaw,
Ye'll wake te find one day
Ye're all washed up, ye've got the push,
Because yer hair's turned grey.
For fifty years Aa've worked in the pits
And mony's the change Aa've seen.
When Aa wes young and a man wes done
They set him on the screens.
But noo today, when yer hair turns grey,
Like a squaddie with his aad peg leg,
They'll tell ye ye're ne use ony mair,
Get oot on the road and beg.

Ee aye, Aa Could Hew

When Aa was young an' in me prime,
Ee aye, Aa could hew,
Wey, Aa was hewin aal the time.
Now me hewin days are through, through,
Now me hewin days are through.

At the face the dust did flee,
Ee aye, Aa could hew,
But now that dust is killin me.
Now me hewin days are through, through,
Now me hewin days are through.

Aa've lain down flat an' shovelled coal,
Ee aye, Aa could hew.
Me eyes did smart in the dust-filled hole.
Now me hewin days are through, through,
Now me hewin days are through.

Aa've worked wi' marras, an' they were men.
Ee aye, Aa could hew.
Wey, they were men an' sons o' men.
Now me hewin days are through, through,
Now me hewin days are through.

It's doon that pit ne mair Aa'll see,
Ee aye, Aa could hew,
But Aa'll carry it round inside o' me,
Now me hewin days are through, through,
Now me hewin days are through.

V

The Perils of Mining

V

THE PERILS OF MINING

Reporting on coal-mining conditions in 1844, Frederick Engels declared: 'In the whole British Empire there is no occupation in which a man may meet his end in so many diverse ways as this one. The coal-mine is the scene of a multitude of the most terrifying calamities, and these come directly from the selfishness of the bourgeoisie. The hydrocarbon gas which develops so freely in these mines, forms, when combined with atmospheric air, an explosive which takes fire upon coming into contact with a flame and kills everyone within its reach. Such explosions take place, in one mine or another, nearly every day . . . If an explosion occurs, the recklessness of the miner is blamed, though the bourgeois might have made the explosion wellnigh impossible by supplying good ventilation.'

Mines are better ventilated now, but explosions and other disasters are not uncommon and they form a great part of the miners' cares. Throughout the course of the mine ballad tradition, probably the most consistent part is concerned with such disasters. Some of these ballads were made by the writers of street songs, who put the sensational news of the day into broadsheet verse; but as often as not it would be a miner-balladeer, moved by the drama and horror of the disaster, who made the commemorative song.

After the eighteen-forties, the standard form for pit disaster ballads came to be the alternating eight- and six-syllable phrases of the Irish street song, the 'come-all-ye', a style of ballad that has proved more adaptable than any other for the requirements of industrial folklore.

Many an old miner preserves the saga of some disaster that has faded even from local memory, such as the elegiac ballad of the explosion at Dixon's Pit, High Blantyre, in 1877, or the song celebrating the four heroic rescue-men of the Donibristle mine in 1901. Another calamity whose memory has been preserved by a pitman-poet is that of the single-shaft Hester pit at Hartley, Northumberland, on January 16th, 1862, when the forty-ton beam of the pumping engine broke and fell down the shaft, blocking it, destroying the ventilation, and entombing 204 men and boys. Joseph Skipsey – whose father had been shot and killed during the 1832 strike – wrote a ballad of the disaster, to be read at meetings in aid of the widows and children. One present at the time reports: 'It was scarcely like reading: [Skipsey] entered so evidently into the spirit of the thing, and brought out the terrible tragic nature of the slow death creeping over father and son – the scenes in some places where he read it were almost too painful.'

Over the last hundred years, the pit-disaster ballads have shown less variation of type than any other kind of miners' songs, as may be seen from the 'classic' run of the song treating of the relatively recent disaster at Easington in 1951.

The Collier's Lament

Almost a week in deep distress within the pit we lay,
Without one ray of cheering hope to pass our time away.
So long we were in want of food, but three were found alive,
And I have been the most distressed of those who did survive.

One May the third in the morning, that fatal roof gave way.
Seven men were underground, abandoned hope that day.
But still we were relieved, as providence would have,
And still I wish with all my heart that this had been my grave.

In a hospital I was received, as you may understand;
But, to my loss, I left there two fingers of my right hand.
This hinders me from following my trade, or getting my infants bread;
And when I hear their pitiful cries, I wish that I was dead.

And so my mournful story is here unto you told.
So helpless is my fate that I freely do unfold.
That I am no impostor, you may freely understand,
It's very true what I tell to you, for look at my right hand.

The Wallsend Explosion

What is this that strikes my ear?
Most awful is the sound I hear.
Listening to the rolling sound,
I hear it comes from underground.

Now the flame and smoke ascend,
The awful shock the structures rend,
Fragments in confusion fly,
The awful scene obscures the sky.

Soon the dreadful news is spread:
The pit has blasted, all are dead.
Now the wives distracted runs,
To seek their husband and their sons.

Each mind is then filled with despair.
Some wring their hands and tear their hair.
The shrieks most dreadful is to hear,
While some more silent drops the tear.

But mature kind give some relief,
A flood of tears gives vent to grief.
In broken accents one do say:
Sad Sabbath for a wedding day.

Another prospect strikes my view:
A group of men all dressed in blue,
And miner slaves do them attend
When down the pit they all descend.

The miners then both far and near,
In numbers on the hills appear,
Enquiring then what must be done
As in obedience swift they run.

Then mark the banksman, with his eye
Upon the rapper fixed close by.
And as it moves, then hear him say:
'There's something on. Come, bend away!'

And now begins the tragic scene.
Wives, mothers, sisters, sweethearts scream,
And when to bank the corpses come,
To claim their own each trembling run.

But when the corpse is viewed, there's none can claim,
Nor yet the mangled body name.
Some buttons they take from the clothes,
And ask if anyone these buttons knows.

The buttons they are passed around,
And soon as owner there is found.
A mother cries: 'Those are my son's!'
To get him in her arms she runs.

But some kind friend stepped in between,
To keep her from the shocking scene,
And lends her arm to take her home,
When to the door the corpse do come.

Now strangers flock to view the scene,
And females help the corpse to clean.
They strip the clothes, wash hands and face,
And still no features they can trace.

The coffin's brought, the corpse put in,
A tuft of hair in paper pin,
And in the drawer is put away
For to look at some other day.

Again unto the pit I turn.
Each object viewed it makes me mourn.
Coffins upon coffins piled,
Some for the father, some the child.

Now down the pit the men are sent
Enough to make the rocks to rend,
And when each body there is found,
It's closely coffined underground.

Take off the shoes, pull off some hose
Or other fragments of the clothes
And give them to the gazing throng,
To see to whom they do belong.

But before all the bodies found,
They see something which pain the mind.
For as they make their way in-bye,
A flame of fire they do espy.

And now they quickly make away.
It would be dangerous to stay.
Then right outbye they run with dread.
The pit's on fire! The news is spread.

And when to bank they all do come,
They then consult what must be done,
And soon they all as one agree
To close the shaft immediately.

Buntons then and planks do come,
And soon the scaffold's fixed and done,
With full four feet of tempered clay,
And water lest it run away.

Then full eight weeks or maybe more
Or they the mine again explore.
And when the pit they do go round,
And all the putrid bodies found,

They're coffined close and sent away,
And each one buried without delay.
And here do end this mournful scene.
The pit is now both safe and clean.

The coroner has verdict passed:
'It was by accident, the blast.
The case I have examined through,
The overman and viewer too,
And all the witnesses declare,
The gas was small, but strong the air.

So gentlemen, you all may see
What was observed by Mr B.
The fact so plain needs no debate;
We all must be subject to fate.

The Scriptures they explicit are,
How God does watch each numbered hair,
And while we such disasters see,
They must all come by Heaven's decree.'

The Trapper Boy's Dream
(or The West Moor Colliery Explosion)

Silence was in the lowly cot, the trapper boy reclined
Upon his clean though humble couch, prepared by his mother kind.
And near him sat the matron form, as fearful to molest
The slumber that she thought partook of dreaming and unrest.

And now he starts, he mutters words breathed from his lips of fear
And drops of sweat upon his brow like shining beads appear.
And now he lifts his trembling hand and holds his burning brow;
'Save me, oh save me,' loudly cries and wakens wildly now.

'Oh, mother, such a dreadful dream! A cup of water, pray.'
How eagerly he takes the draught, his cheek as cold as clay.
'Oh, what a dream! I would not sleep again to see that sight
For all the gold the earth contains – oh, horror and affright!

'I thought that at my post I sat, upon my duty bent,
When suddenly there came a sound as if the mine was rent:
And then the earth rocked to and fro, and I strove for help to call
For o'er my head a mass of coal hung ready for to fall.

'It swayed and tottered, still it hung, as held by secret power,
And as I gazed such horrid faces round me seemed to lower.
Grim demons looked with scowling eye, and nearer then they came.
They smote and dashed me to the earth and turned my heart to flame.

'I felt my bones were ground to dust, my carcase mangled lay.
My head I found was crushed, the brains were mingled with the clay,
And yet the state of sense was left, the terror of the heat
Was vivid, and I thought how blest if I from life could part.

'I sat as it were without my eyes, and reasoned without brains,
I felt a death that dying man ne'er felt in worst of pains.
And without tongue a voice I found, and called on God to save.
Oh, mother, did I ever think so earnestly to crave.

‘God heard me, and I dream no more. But do not let me go
Again to the dreary mine. I’m warned to shun the blow.
You’re weeping, mother, and I know what you would wish to say,
That I am foolish to regard a dream that fades away.

‘Oh, do not weep, and I will chase this vision from my brain.
’Tis for you, mother; we are poor, and I’ll go to work again.’
It’s done – and now the shapeless body rests on the same bed
Where whiles it dreamt, from which it rose, on which it now lies dead.
Proclaiming to the selfish world that labours to destroy
The better feelings of the poor – this was a trapper boy!

The Jarrow Pit Explosion

For these poor colliers drop a tear
Whose mournful fate I now declare;
They were cut off in youth and bloom,
And laid in an untimely tomb.

They were full of health and free from care,
They little thought their end so near;
In the midst of life we are in death,
The deed then stopp'd their precious breath.

Hark! Hark! what means that dreadful sound
Which spreads dismay and fear around,
Alas! it from the pit doth come,
It's sealed the unfortunate collier's doom.

Now parents cry in accents wild:
Alas, thou art gone my much lov'd child,
Oh that I could have died to save
My son from this untimely grave.

The widow's heart with grief is torn
Orphans for their father mourn,
Their tender hearts are nearly broke
By this sudden awful, awful stroke.

'Twas a doleful sight to see
Their lifeless bodies as they lay,
Some in their youth, some men full grown,
Grim death has claimed them for his own.

When these poor colliers were conveyed
And in the grave their bodies laid,
Some thousands did assemble there
And for their fate shed many a tear.

Their glass is run, their day is past,
No more they feel life's stormy blast,
We hope they've reached yon peaceful shore
Where pain and sorrow are felt no more.

The St Helens Colliery Explosion

Attend a while, ye Christians, that in the nation dwell,
Unto a fatal accident which lately has befell.
Few men of all the sons of earth, wherever they may be found,
In greater peril lie than those who work beneath the ground.

> Christians, send your prayers on high
> For those that in Sofford Colliery died.

A sad heart-rending catastrophe to you I will unfold,
For such a dreadful tale of woe before was never told.
'Twas on the twenty-third of June, eighteen hundred and fortynine,
When seven poor colliers lost their lives at the Sofford coal mine.

It makes me shudder when I think on their dangers dark and dire,
Of damps, foul air and falling earth, of water and of fire.
But let me tell my woeful tale, and reader, keep in view
Through every line and every verse that what I state is true.

It was at Sofford Colliery this sad event befell,
Near to the town of St Helens, by hundreds known full well,
It was on Saturday the twenty-third, sad news through St Helens spread
That with firedamp the pit was filled, and seven poor colliers dead.

There were sixty miners at the time working underground.
Twenty of them were sorely bruised and seven dead were found.
To see their wives and children it would melt a heart of stone,
Some looking for a husband, some a father or a son.

Fine healthy men in youth and bloom by death were snatched away,
All by the sad explosion upon that fatal day,
While these mournful widows their hearts did ache with pain
To think that morn would be the last they'd ever meet again.

Children are now left fatherless all through this sad event.
Sad tears in torrents they did fall, while hundreds did lament
For friends and their relations who met their awful doom.
Let's hope their souls are now at rest, while in the silent tomb.

The Fatal Coal-Pit Explosion at Nitshill

Oh, sad is the news in the village of Nitshill,
The bright joyous days for a long time has gone;
Now naethin' is heard but sad weepin' and moanin'
For husbands, and fathers, and brothers and sons.
The year fifty-one we long will remember;
The 15th of March was the sad fatal day, .
When early that mornin', and no danger dreadin',
Sixty-three brave coal-miners to their work hied away.

But, sad to relate about five that same mornin'
A dreadful explosion was heard through the air,
And columns of fire from the pit-mouth ascended,
When wives and relatives to the spot did repair.
Oh, no human pen can describe the sad anguish
That filled their sad heart, their cries rent the air;
Both fathers and mothers, and wives and their children,
Each bosom was fill'd with sore grief and despair.

When the shock was first felt round the village of Nitshill,
An old aged parent came forth in despair;
He utter'd no cries, to the pit-mouth he bounded,
Crying, where is my three sons? Oh, tell me! Oh, where?
This poor man was frantic, with overpower'd anguish,
My sons now I'll follow to death! he did cry,
Then made to jump down as the fire it ascended,
And was only prevented by those standing by.

Oh, pity, good people, that poor distress'd woman
That lost her four sons, and her husband so dear.
In sad grief and anguish now she's daily declinin',
No one to console her, nor son for to cheer.
And oft she exclaims when on bed she's reclinin',
Come Death, quick and free me of my grief and woe!
There's nought in this world can e'er give me pleasure;
To the grave to my husband and children I'll go.'

Draw near, ye brave miners, and coal-masters of Scotia,
   Give ear to the widow and poor orphan's cry,
And give your assistance to shield and protect them,
   And you will be rewarded by the ruler on high.
To us all death is sure though the time is uncertain,
   No one 'gainst its shafts was e'er known to prevail,
Sixty-three that were well upon that fatal morning,
   Only two are now left for to tell the sad tale.

The Late Colliery Explosion at Patricroft, Wigan

You feeling Christians one and all, of high and low degree,
I pray you give attention and listen unto me;
The fate of these poor colliers in sorrow to lament,
Who suddenly has lost their lives by this coal-pit accident.

Upon the 20th day of February these men and boys we're told
Went to their labour so fearlessly and bold,
Never thinking sudden death would strike them down so soon,
But oh, alas! the fire blast surrounded them at noon.

The news of this explosion spread through the country round,
While numbers of poor colliers were suffering under ground
Ascending prayers to heaven in the suffocating blast,
While others lay in agony, and breathing of the last.

As soon as it was possible, assistance did go down,
To their surprise, both men and boys, alive and dead was found,
With mingled flesh and broken bones, and all to pieces rent,
No tongues can tell the sufferings those creatures underwent.

Oh, what a dreadful spectacle for to behold the dead,
When all their lifeless bodies upon the bank was laid.
To recognize their features, it scarcely could be done,
The uncle from the cousin, or the father from the son.

They were so mutilated, alas! it is well known,
How some of these poor colliers were all to pieces blown,
The head from off the shoulders was severed I declare,
While bodies, legs and arms, they were scattered here and there.

Friends and relations gathered round to recognize their own.
The cries and lamentations would melt a heart of stone.
The loving wives and children, with sorrow in each face,
While death and desolation in the pit was taking place.

No one can tell the hardship poor colliers undergo,
Or dangers they encounter when they are down below.
They know not when they once go down but it may be their last,
May God above preserve them from damp and fiery blast.

Oh, never yet in Wigan such a sight has not been known,
Which causes many thousands in sorrow to bemoan.
The fate of these poor colliers, it will be handed down,
To the rising generation, for many miles around.

May God look down from his almighty throne,
Upon those orphan children, who now are left alone,
While many a loving mother dear do sorely feel the smart,
May God look down upon them, and soothe their aching heart.

May those surviving colliers who pain and sorrrow felt,
Be soon restored one and all, unto their perfect health,
And those that lie in slumber and numbered with the dead,
Oh, let us hope, good Christians, their souls to heaven are fled.

The Cummer Colliery Explosion

You feeling Christians, give attention,
Listen to my mournful tale.
'Tis of a dreadful accident,
The truth to you I will reveal.
'Tis of one hundred and ten colliers,
Who did appear in health and glee,
By fire-damp they were cut off,
And sent into eternity.

Colliers all pray take a lesson.
Let their fate a warning be.
Pray to God, both night and morning,
From all explosions to keep you free.

'Twas on July the fifteenth day,
Now the truth I will make known,
And when you hear these verses through,
'Tis enough to melt a heart of stone!
Soon as this sad news was spread,
People flocked from all around,
To see these poor unfortunate dead
Colliers brought up from underground.

The shrieks and cries did rend the air,
I am sorry for to say,
Of fathers, mothers, wives, and children,
As to the pit they bent their way.
To see their mangled bodies brought up,
It is most awful to relate,
Some could scarcely know their own,
They were burnt to such a state.

To see one hundred and ten colliers,
  Lying dead upon the ground,
Oh, what sensation it did cause,
  Now in all the country round!
They little thought that very morning
  As they were rising from their bed,
So soon they would be listless clay,
  And be numbered with the dead.

May God above protect their widows,
  And the children fatherless,
Of those poor unfortunate colliers,
  Who are now in deep distress!
Lamenting for their loving husbands,
  And mothers for their sons we're told,
By fire-damp they were cut off,
  It was most shocking to behold.

Oh, God above, look on my children,
  Fathers and mothers aloud did cry,
And wives, their screams did rend the air,
  And for their husbands they did sigh.
They're gone a little time before us,
  And I hope to endless rest.
May they sit at God's right hand,
  And there be numbered with the blest.

The Hartley Calamity

The Hartley men are noble, and ye'll hear a tale of woe,
I'll tell the doom of the Hartley men, the year of sixty-two.
'Twas on a Thursday morning in the first month of the year,
When there befell the thing that well may rend the heart to hear.

Ere chantecleer with music rare awakes the old homestead,
The Hartley men are up and off to earn their daily bread.
On, on they toil, with heat they broil, and streams of sweat do glue
The stour into their skins till they are black as the coals they hew.

Now to and fro the putters go, the waggons to and fro,
And clang on clang of wheel and hoof ring in the mine below.
The din and strife of human life awakes in 'wall' and 'board',
When all at once a shock is felt which makes each heart-beat heard.

Each bosom thuds as each his duds then snatches and away,
And to the distant shaft he flees with all the speed he may.
Each, all they flee by two, by three, they seek the shaft to seek
An answer in each other's face to what they may not speak.

Are we entombed? they seem to ask, for the shaft is closed, and no
Escape have they to God's bright day from out the night below.
So stand in pain the Hartley men, and swiftly o'er them comes
The memory of home and all that binds us to our homes.

Despair at length renews their strength, and they the shaft must clear,
And soon the sound of mall and pick half drowns the voice of fear,
And hark, to the sound of mall below, do sounds above reply?
Hurrah, hurrah for the Hartley men, for now their rescue's nigh.

Their rescue nigh? The sounds of joy and hope have ceased, and ere
A breath is drawn, a rumble's heard that drives them to despair.
Together now behold them bow, their burdened souls unload
In cries that never rise in vain unto the living God.

Whilst yet they kneel, again they feel their strength renewed again,
For the swing and ring of the mall attest the might of Hartley men.
And hark, to the blow of the mall below, do sounds above reply?
Hurrah, hurrah for the Hartley men, for now their rescue's nigh!

But now that light, erewhile so bright, no longer lights the scene,
A cloud of mist the light has kissed and shorn it of its sheen.
A cloud of mist the light has kissed, and see, along does crawl,
Till one by one the lights are smote, and darkness covers all.

'Oh father, till the shaft is cleared, close, close beside me keep.
My eyelids are together glued, and I – and I must sleep.'
'Sleep, darling, sleep and I will keep close by – heigh ho – ' To keep
Himself awake the father strives, but soon he too must sleep.

And fathers and mothers and sisters and brothers, the lover, the new-made bride,
A vigil kept for those who slept from eve to morning tide.
But they slept, still sleep, in silence dread, two hundred old and young,
To awake when heaven and earth have sped and the last dread trumpet sung.

The Oaks Pit Explosion

There is weeping and wailing in many a dwelling,
Where happiness reigned but a short time ago.
For friends and relations, there's numbers left mourning,
In accents of sadness, in accents of woe.
Oh, who can console now bereaved and afflicted,
Or comfort the orphans who silent deplore,
O'er the cheeks of the widow the tear is fast falling,
Her son, her support, he alas is no more.

To the Oaks Pit near Barnsley on the 12th of December
About four hundred colliers to work they did go,
In health and in strength upon that fatal morning,
Not thinking of danger they descend below.
But that afternoon without one moment's warning,
That coal pit exploded with a terrific roar.
Three hundred and forty eight by that fearful explosion,
Their lives they have lost, which we deeply deplore.

Of this dreadful disaster the news spread like lightning.
And the people came running from both far and near,
And the widows and orphans their cries were heartrending.
As they called on their husbands and fathers so dear,
While brothers and sisters in groups were seen mourning.
And old aged parents their gray locks they tore
And maidens lamented the fate of their lovers,
By that fatal explosion they'll ne'er see them more.

Of those brave volunteers who so boldly did venture,
Down that ill fated pit the poor sufferers to save,
They did all lose their lives by the second explosion,
Except Samuel Brown, that young hero so brave.
After twenty long hours in that gloomy sepulchre,
He got to the shaft and for help did implore.
Mr Embelton and Mammott went down and they found him,
And brave Brown in life to the surface they bore.

That direfull explosion at the Oaks pit near Barnsley,
  O'er the mining districts has cast a sad gloom.
Of four hundred colliers in the Oaks pit that morning.
  Now three hundred and forty eight does lie in the tomb,
They have left aged parents with none to befriend them,
  Whose pleasures on earth it for ever has fled,
And who now will comfort the widows and orphans
  Who mourns for their friends that now sleeps with the dead.

The Starlaw Disaster

It was eighteen and seventy, Aprile the ninth day,
Ninety six men and boys for their work took their way.
In health and in strength down the shaft they did go,
Never dreaming of how many would lie low.

For about twelve o'clock on that same fatal day
'The pit shaft's on fire,' the roadsman did say.
And quick through the workins the alarum he gave,
All praying to their maker their sweet lives to save.

All eagerly ran to get on to the cage,
But the fire in the shaft like a furnace did rage.
All praise to young John Steel who at the engine did stand,
And forty-eight safe on the bank he did land.

William Ralston, William Rushford and young David Muir,
By that terrible disaster you'll see them no more.
Patrick and Peter M'Comiskie, aye, and Swanson likewise
By that terrible disaster in their cold grave now lies.

Now widows and orphans who are now left to mourn,
By that awful disaster they will never return;
But God is so merciful as all mankind knows,
He will share in their sorrow and soften their woes.

Melancholy Explosion at the Moss Pit

In those villages of England
Where our coal pits do abound,
We oftimes hear of dire explosions,
Spreading grief for miles around;
Happy wives are soon made widows,
Children lose their only stay,
Aged fathers, dear old mothers
Think on many a fatal day.

Down a deep and fiery coal mine
Seventy colliers burning lie,
And 'twas while attempting rescue,
Many more were doomed to die.

In the deep gloom of the coal mine,
Hardy men are working there,
They cannot see the smiling sunshine,
Cannot breathe the fragrant air;
When the fiery damp comes o'er them,
Then grim death is lurking near.
They meet the fate of those poor colliers
At the Moss Pit, Lancashire.

On the sixth day of September,
Seventy men were struck with death.
That fearful day we all remember,
The effect of fire damp's scorching breath.
'Twas found there was no hope of saving;
All had met a fearful doom,
Although they tried, all dangers braving,
None could reach their fiery tomb.

When they thought the dangers over,
Men stood round of noted skill,
And the pit's mouth unclosed,
With sweet air the workers fill;
When quick without a moment's warning,
Not even one despairing cry,
Their bodies were blown to the winds of morning,
The flames and smoke rose mountains high.

Soon poor men and women assembled,
And in anguish met around.
Gallant men stood there and trembled
At that second fatal sound.
The foreman, Walsh, whose sons had perished,
In the pit they're lying now,
Trying to save those he had cherished
Met his death on the pit's brow.

Oh tell me, tell me, dearest mother,
Is my father with the slain?
Is my only little brother
Coming from the pit again?
When will father be returning,
To join me in my childish play?
The mother cries, with anguish burning
Your father is in heaven, I pray.

The Pit Accident at Pelsall

Oh! list all you Christians as you're passing by,
It's about those poor colliers, a mother and boy,
For all was commotion at every street door,
They rushed to Hall Collery their hearts full and sore.
Some they are running, they cry and they shout,
'Our father is killed at Pelsall I doubt,
The water broke on them, took them in surprise,
And stopped ventilation which cost them their lives.'

'Oh! do not weep, mother, our father will come,
For we are so happy when he is at home,
He is a poor collier, I can't make it out yet,
The reason he's stopping so long at the pit.
At six in the evening from his work he would go,
And we was so happy when his face he would show,
He'd call us his darling, play with us awhile,
So don't weep, dear mother, but give us a smile.'

'Hush, do not murmur, a secret I'll tell,
Your father an angel in heaven doth dwell.
Although he was loving to his wife and child,
He'll not be returning, the thought drives me wild.
It's true he's a collier with twenty-one more,
Have gone to the angels on that distant shore,
So that is the reason that I weep and cry,
He has left me a widow, you an orphan boy.

He worked at Pelsall the place that you know.
You went with his dinner a short time ago,
Now you'll never ramble to that pit after him.'
So think of the colliers and poor little Jim.
So all take a warning and do not long wait,
For death is a-lurking close to our gate,
The hour and the minute we never can tell,
That we shall be summon'd above for to dwell.

The Blantyre Explosion (I)

By Clyde's bonny banks where I sadly did wander,
Among the pit-heaps as evening drew nigh,
I spied a fair maiden all dressed in deep mourning,
A-weeping and wailing with many a sigh.

I stepped up beside her and thus I addressed her:
'Pray tell me, fair maiden, of your trouble and pain.'
Sobbing and sighing, at last she did answer:
'Johnny Murphy, kind sir, was my true lover's name.

'Twenty-one years of age, full of youth and good-looking,
To work down the mines from High Blantyre he came.
The wedding was fixed, all the guests was invited.
That calm summer evening young Johnny was slain.

'The explosion was heard, all the women and children
With pale anxious faces they haste to the mine.
When the truth was made known, the hills rang with their moaning.
Three hundred and ten young miners were slain.'

Now husbands and wives and sweethearts and brothers,
That Blantyre explosion they'll never forget.
And all you young miners that hear my sad story,
Shed a tear for the victims who're laid to their rest.

The Blantyre Explosion (II)

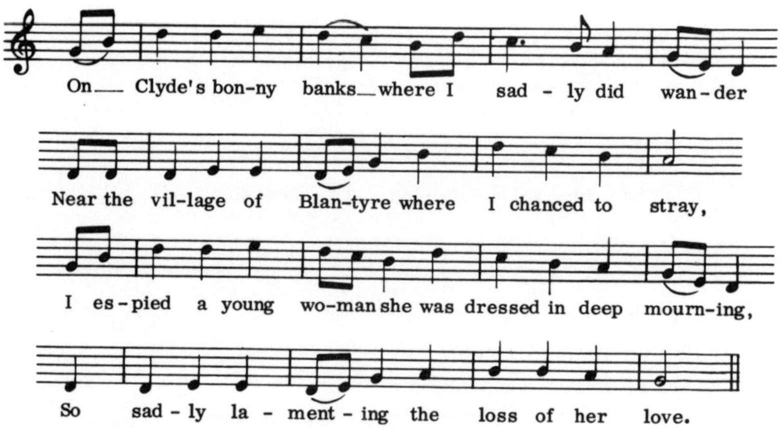

On Clyde's bonny banks where I lately did wander,
Near the village of Blantyre where I chanced to stray,
I espied a young woman, she was dressed in deep mourning
So sadly lamenting the fate of her love.

I boldly stepped to her; said I, My poor woman,
Come tell me the cause of your trouble and woe.
I do hear you lamenting the fate of some young man,
His name and what happened him I'd like for to know.

With sighing and sobbing she at length then made answer,
John Murphy, kind sir, was my true lover's name,
Twenty-one years of age and a mild good behaviour,
To work in the mines of High Blantyre he came.

On the eleventh of December I long will remember,
In health and in strength to his labour did go,
But on that fatal morning without one moment's warning,
Two hundred and ten in cold death did lie low.

There was fathers and mothers, there was widows and orphans,
In Stonefield, High Blantyre, where hundreds do mourn,
Ah, there was old aged parents for their sons they loved dearly,
By that sad explosion will never return.

But the spring it will come with the flowers of summer,
That blooms through its wildness so lovely and fair.
I will gather the snowdrops, primroses and daisies,
Round my true lover's grave, I will transplant them there.

For they say it's not right for the dead to be grieved,
There's nothing but trouble bestowed on me.
He is gone from this world but a short time before me,
In hopes to rejoin him in sweet purity.

Fearful Colliery Explosion in Scotland

A dreadful and heartrendering sight in Scotland has been seen,
Where far away from friends and light poor men destroyed have been.
When working down beneath the ground, toiling for their bread,
About two hundred living souls were numbered with the dead.

The men were hard at work we find in Dixon pit that day,
Gallant hearts, both good and kind, God help them now, we say.
Without a moment's warning then the fatal firedamp came,
And about two hundred boys and men were killed by the deadly flame.

Hundreds stood around the pit in solemn silence there.
By the winter's sun each face was lit, but only showed despair,
Mourning for some missing one they never more would see,
Until the pathway they have trod of dark eternity.

The Trimdon Grange Explosion

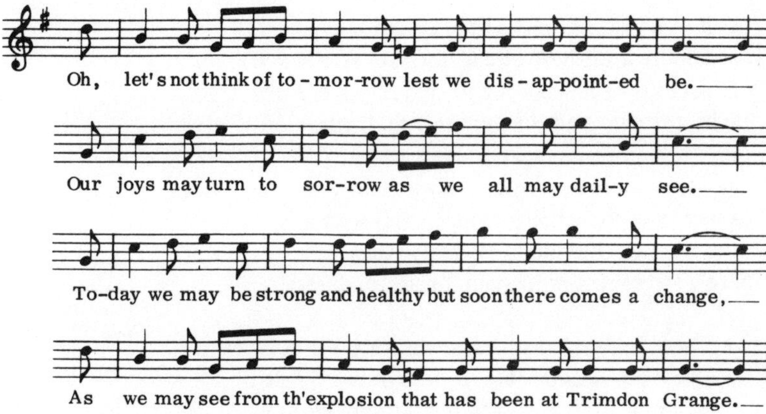

Oh, let's not think of tomorrow, lest we disappointed be.
Our joys may turn to sorrow as we all may daily see.
Today we may be strong and healthy, but soon there comes a change,
As we may learn from the explosion that has been at Trimdon Grange.

Men and boys left home that morning for to earn their daily bread,
Nor thought before that evening they'd be numbered with the dead.
Let's think of Mrs Burnett, once had sons but now has none –
By the Trimdon Grange explosion, Joseph, George and James are gone.

February left behind it what will never be forgot;
Weeping widows, helpless children may be found in many a cot.
Now they ask if father's left them, and the mother hangs her head,
With a weeping widow's feelings, tells the child its father's dead.

God protect the lonely widow, help to raise each drooping head.
Be a father to the orphans, never let them cry for bread.
Death will pay us all a visit, they have only gone before.
We'll meet the Trimdon victims where explosions are no more.

Terrible Colliery Explosion at Clifton Hall Pit

We all of us stand on the edge of the grave,
Tossed here and there like a storm-beaten wave.
We don't know a second and scarcely a breath,
One hour in life and the next one in death.

How often we hear of men losing their lives,
Leaving in sorrow their children and wives,
But as sure as the tide to Eternity rolls,
There's sweet rest in heaven for poor colliers' souls.

> There is sweet rest in heaven for poor colliers' souls.
> Their life is uncertain as time onward rolls.
> In life or in death, God bless them, we say.
> Near two hundred colliers have now passed away.

In the village of Pendlebury some collieries stand,
As safe, 'twas supposed, as any in the land.
For several years they had been free from harm,
No reason for trouble, no cause for alarm.

But death comes so silent down in the dark mine.
In the bowels of the earth, the sun cannot shine.
In the Clifton Hall Mine, men working for bread
In a moment were laid, disfigured and dead.

Nearly four hundred poor souls went down on that day,
Little dreaming their lives would be taken away,
But when the explosion seized the men there,
Death came upon them and found them in prayer.

Praying to God to save their dear lives,
Praying for their parents, children and wives.
When their friends saw them, with sorrowful eyes,
The bodies before them they could scarce recognize.

Almost to pieces their bodies were blown.
In terrible forms their bodies were thrown.
Brave explorers by their duties were led
To succour the living and recover the dead.

Down in the mine went many a brave man,
Risking their lives, deny it who can.
They could not restore dear life to the dead,
But only prepare them for their last lonely bed.

'Tis frightful to think of this loss of life,
The pity of the orphans, each mother and wife.
The face of the poor ones they will see no more
Till eternity joins them on a happier shore.

Destroyed in the midst of mence, health and prime,
They will be rewarded in Heaven's good time.
From all that they loved they've been taken away,
But Heaven is their home 'midst the angels today.

mence = pride

The Donibristle Mossburn Disaster

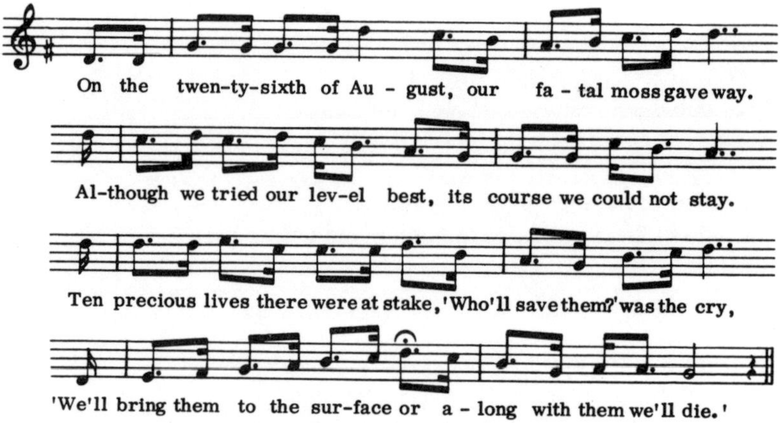

On the twenty-sixth of August, our fatal moss gave way.
Although we tried our level best, its course we could not stay.
Ten precious lives there were at stake. 'Who'll save them?' was the cry.
'We'll bring them to the surface or along with them we'll die.'

> There was Rattray and McDonald, Hynd and Paterson,
> Too well they knew the danger and the risk they had to run.
> They never stopped to count the cost. 'We'll save them', was the cry.
> 'We'll bring them to the surface or along with them we'll die.'

They stepped upon the cage, they were ready for the fray.
They all meant business as they belled themselves away.
Soon they reached the bottom, far from the light of day,
And went to search the workings, and Tom Rattray led the way.

They lost their lives, God help them. Ah yes, it was a fact.
Someone put in a stopping and they never did get back.
Was that not another blunder? My God, it was a sin.
To put a stopping where they did, it closed our heroes in.

We never shall forget them, though they have lost their lives,
So let us pay attention to their children and their wives.
It simply is our duty now, and let us all beware.
Their fathers died a noble death and left them in our care.

There was Rattray and McDonald, Hynd and Paterson,
Too well they knew the danger and the risk they had to run.
They never stopped to count the cost. 'We'll save them', was the cry.
'We'll bring them to the surface or along with them we'll die.'

The Maypole Disaster

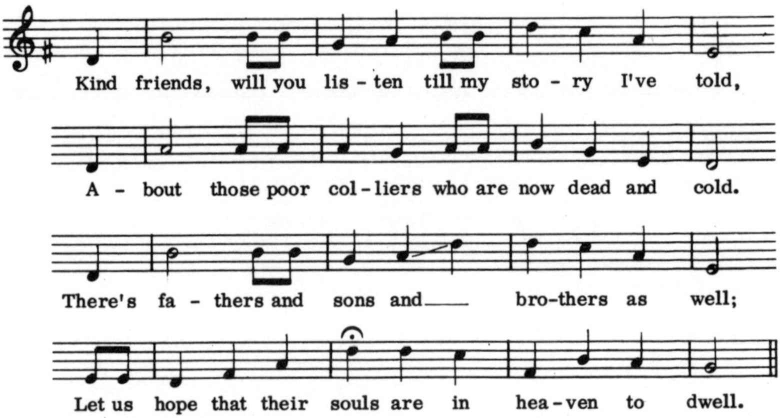

Kind friends, will you listen till my story I've told,
About those poor colliers who are now dead and cold.
There's fathers and sons, and brothers as well;
Let us hope that their souls are in heaven to dwell.

There's widows and sisters who are called for to groan.
There's many sweethearts now left for to mourn,
Not thinking so soon from their loved ones to part,
The widows are left with a sad broken heart.

Eighteenth of August the explosion occurred.
For many miles round it was felt and was heard.
The scenes at the pithead would fairly make you stare,
To see grieving widows with broken hearts there.

There's one little lad, to the policeman he said:
'Please sir, let me through, for my daddy is there.'
With tears in his eyes, the policeman said: 'Lad,
You can't do any good here; go home and pray for your dad.'

There's mothers and sisters and sweethearts in tears,
Stood watching and waiting in sadness and fear,
All night in the cold till the break of the day.,
Before they attempted to go on their way.

Then off with your hats to the lads that win coal,
And the lives that were lost in that pit called Maypole.
But we never know when death comes our way,
These are the last words we've all got to say:

They are gone from this world with its sorrows and cares,
Gone to their Maker, his glory to share.
They are all gone to rest, along with the best.
May their souls rest in peace; may God answer their prayer.

The Maypole Pit Disaster

Our hearts are sad today for the miners gone away
By death, while working in the Maypole mine.
As each went his task to do in that pit, they little knew
That ne'er again they'd see the bright sunshine.

As the day was wearing on and warning came to none,
When the firedamp spread around its awful breath.
Then came a sudden roar; in a flash their lives was o'er,
They lay all wrapped within the arms of death.

The news soon spread around, and willing hearts were found,
To start the work of rescue down below.
All honour to their names; to face those fiery flames,
For their fellow workmen's sake those heroes go.

The deadly damp they fought and through awful perils on,
With aching hearts, their comrades' lives to save.
But at last all hope was past to face the fiery blast,
They found the Maypole pit had made the miners' grave.

With faces drawn and white, all through that dreadful night,
Mothers, wives and sisters waited there,
Friends and relations too, and when the truth they knew,
Their hearts were filled with grief and wild despair.

All England will regret the awful fate they met,
We sorrow for all the loved ones left behind.
They are desolate indeed, let us help them in their need,
And cheer those helpless ones with friendship kind.

The Gresford Disaster

You've heard of the Gresford disaster,
And the terrible price that was paid.
Two hundred and forty-two colliers were lost,
And three men of a rescue brigade.

It occurred in the month of September,
At three in the morning, that pit
Was racked by a violent explosion
In the Dennis where gas lay so thick.

The gas in the Dennis deep section
Was packed there like snow in a drift,
And many a man had to leave the coal-face
Before he had worked out his shift.

A fortnight before the explosion,
To the shotfirer, Tomlinson cried:
'If you fire that shot we'll be all blown to hell!'
And no one can say that he lied.

The fireman's reports they are missing,
The records of forty-two days;
The colliery manager had them destroyed
To cover his criminal ways.

Down there in the dark they are lying,
They died for nine shillings a day.
They have worked out their shift and now they must lie
In the darkness until Judgement Day.

The Lord Mayor of London's collecting
To help both our children and wives.
The owners have sent some white lilies
To pay for the poor colliers' lives.

Farewell, our dear wives and our children,
Farewell, our old comrades as well.
Don't send your sons down the dark dreary pit;
They'll be damned like the sinners in hell.

The Easington Explosion

Come, listen all ye miner lads that tak the road inbye,
And I will tell a tragic tale, when ten times eight did die.
At Easington in 'fifty-one they saw the gates of hell,
But they that lived to see the sight did not live to tell.

The day, the twenty-ninth o' May, a tragic trick of fate
Found the night-shift on the face, the foreshift in the gate.
Explosion wrecked the Quarter seam and killed them all but one,
And many a miner's humble home lost father, brother, son.

'Twas fire-damp beneath the cut, coal-dust fed the flame
That roared outbye till it was spent, then roared inbye again,
Tearing, twisting road supports from their truly bed,
And left behind our sorrow with the dying and the dead.

Within the hour, from Houghton-le the rescue team did come.
With them, hope to find the blind and listen for the dumb;
But hope was killed by after-damp, that gas we miners dread.
Two rescue-men with yellow birds were numbered with the dead.

Time has chained the widows' tears, and hushed the orphans' cry.
For some the gloom will always live, for some the echoes die.
She said: 'God bring them comfort'. Her words I do recall.
No medals are made for miners, but they are heroes all.

The Auchengeich Disaster

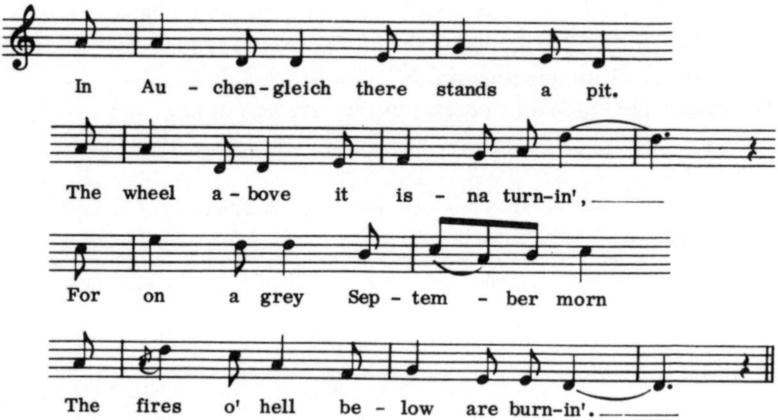

In Auchengeich there stands a pit.
The wheel above it isna turnin',
For on a grey September morn
The fires o' hell below are burnin'.

Though in below the coal lay rich,
It's richer noo for all that burnin';
For forty-seven brave men are deid,
Tae wives an' sweethearts nane returnin'.

The seams are thick in Auchengeich.
The coal below is black an' glistenin'.
But the cost o' coal is far ower dear;
For human lives there is nae reckonin'.

For coal is black an' coal is red,
An' coal is rich beyond a treasure.
It's black wi' work an' red wi' blood;
Its richness noo in lives we measure.

Far better that we'd never wrought
A thousand years o' work an' grievin'.
The coal is black like the mournin' shroud
The women left behind they're weavin'.

In Auchengeich there stands a pit.
The wheel above it isna turnin',
For on a grey September morn
The fires o' hell below are burnin'.

The Watter's Brokken In

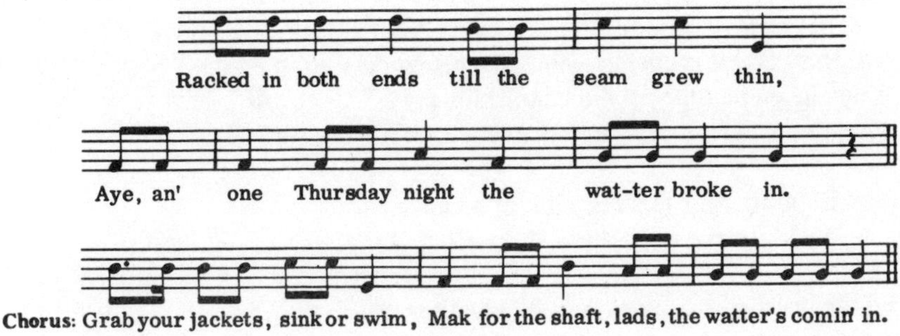

Oh, Fishburn Colliery, December ’sixty-eight,
The trepanners were gannin champion, they never stopped te get their bait,
An’ Number Twenty in the Harvey seam
Was standin as well as a face has ever been.

In comes the under-manager, looked aroond an’ said:
‘Get out, me lads, ye’re ower near the limestone bed.’
But the gaffers at Greenlane said: ‘Nowt o’ the kind;
Harvey coal’s ower good te leave behind.’

So they ploughed away, they took big chances,
Cut the stalls an’ fired the canches,
Racked in both ends till the seam grew thin;
Aye, an’ one Thursday night the watter comin’ in.

Grab your jackets, sink or swim,
Mak’ for the shaft, lads, the watter’s brokken in.

Noo, the Harvey developments were the first te go.
They’d run down an’ drownded Ronnie Pears’s Imeco.
Flooded out the Busty an’ the loco road.
Aa says: ‘Aa’ll bet somebody gets the whole toad.’

Grab your jackets, sink or swim,
Mak’ for the shaft, lads, the watter’s brokken in.

The manager sent for his deputy, says: 'Gan an' tell us what we've got.'
An' John Hurst grabs his oilyskins, his waders an' the lot.
When he came outbye, Aa heard him say:
'Why, you canna see much for the watter an' the spray.'

So grab your jackets, sink or swim,
Mak' for the shaft, lads, the watter's brokken in.

So they rushed big pipes fre Philadelph-i-ay,
An' they run them aal out along the loco way.
Built dams, got pumps, did aal they could.
It took three weeks te stem that flood.

An' grab your jackets, sink or swim,
Mak' for the shaft, lads, the watter's brokken in.

Wey, this story has a moral, an' the moral is this:
Even though ye work at Greenlane, don't take extra risks;
For it might be very nice gettin a bit extra coal,
But it's not nice sittin on a black-an'-blue . . . bot.

Grab your jackets, sink or swim,
We're aal at the shaft, lads, the watter's brokken in.

Johnny Seddon

As I went out in summertime all for to take the air,
I met a handsome maiden down by the river clear.
She wept and she lamented and bitterly she cried,
Saying: My curse upon the cruel mine where Johnny Seddon died.

My love he was a collier lad, he worked beneath the ground.
His modest mild behaviour, his equal can't be found.
His two bright eyes and yellow hair, his cheeks a rosy red,
But alas, my handsome collier lad lies numbered with the dead.

Last night as I lay on my bed I fell into a dream.
I dreamed a voice came unto me and called me by my name,
Saying: 'Jeannie, lovely Jeannie, for me you need not mourn,
But the cruel stone does crush my bones and I'll never more return.'

Early the next morning, my dream was clarified.
My neighbours all came rushing in. 'John Seddon's dead!' they cried.
'As he was at his work last night, the roof all on him fell.'
The grief and sorrow of my heart no mortal tongue can tell.

Come all you pretty fair maids, I hope you'll lend an ear.
For the grief and sorrow of my heart is more than I can bear.
For once I loved a collier lad, and he loved me also,
But by a fatal accident, he in his grave lies low.

The Miner Boy

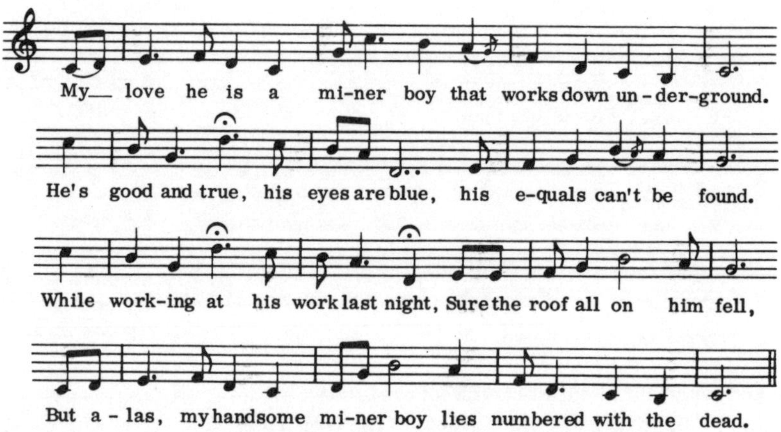

My love he is a miner boy that works down under ground.
He's good and true, his eyes are blue, his equals can't be found.
While working at his work last night, sure, the roof all on him fell,
But alas, my handsome miner boy lies numbered with the dead.

The very night my love was lost I was lying on my bed.
I dreamed a voice came up to me, saying: 'John Sneddon he is dead.'
I dreamed his head lay on my breast as the tears came rolling down,
But I lost my handsome miner boy that his equals can't be found.

The ring was got and the day was set when married we should be.
My love and I was going to sail unto Amerikee.
My love and I was going to sail unto some foreign shore.
For the loss of my handsome miner boy I'm afraid I'll never see more.

The moon shows through my windowpane as the stars do brightly shine.
Summer it will come again when all nature does combine.
When every lass can sport and play, but I his death must mourn,
For the sake of my handsome miner boy; I'm afraid he'll never return.

The Lover's Lament for John Sneddon, a Collier

Now all you pretty maids, I hope you'll lend an ear,
And the cause of all my sorrow you very soon shall hear.
I loved a handsome collier, and he loved me also,
But by a fatal accident in the grave he now lies low.

My love he was a collier braw, and worked beneath the ground.
For mild and good behaviour his equal can't be found.
He had light blue eyes and yellow hair and cheeks like roses red,
But now my handsome collier lad is numbered with the dead.

That night my truelove he was killed, as I lay on my bed,
I thought my love appeared to me and thus to me he cried:
Farewell my dearest Jennie, but for me do not mourn
My spirit's fled in with the dead, I'll never more return.

Then a light broke in my window like the moon when shining clear.
I thought I saw my truelove down the shaft to disappear,
He had his hand upon his head where blood flowed like a stream,
I screamed aloud, I woke in fright and found it was a dream.

But early the next morning then my dream was verified,
The neighbours they came running in. John Sneddon's killed, they cried.
That fatal night while at his work the roof upon him fell.
Now the grief and sorrow at my heart in short there's none can tell.

The banns were out, the day was fixed that married we should be.
My love and I was going to sail unto America,
In hope a fortune for to make upon that fertile shore,
But alas my dearest collier lad I'll never see more.

But the summer will return again and nature will be gay.
The small birds they will sweetly sing and lambs will sport and play.
While other maids will happy be till death I'll constant mourn
For the sake of my poor collier lad who'll never more return.

The Miner's Doom

At five in the morning, as jovial as any,
The miner does rise to his day's work to go.
He kisses his wife and his children so dearly,
And bids them adieu before closing the door.
He goes down the deep shaft with the speed of an arrow,
His heart light and gay without fear or dread.
He's no thought in descending of the danger and peril,
That his life is depending on one single thread.

Now his wife is his queen, his home is his palace,
His children his glory, to maintain with pride.
He'll work like a hero and face every danger,
He'll deprive his own self for their bare feet to hide.
Now his day's work is over, he's homeward returning,
Not thinking the change in one hour could be.
But he thinks how his wife and his children will greet him,
But his wife nor his children he'll nevermore see.

Now his wife had been dreaming of her husband so dearly,
She sees him in danger. 'God help me,' she cried.
So true was her dream to that poor woman's sorrow,
The rope broke while ascending and her dear husband died.
Their home that morning was as jovial as any,
When a dark cloud came rolling straight over the door.
A widow, three children were then left to mourn him,
The one that they never will see any more.

The day of his funeral, a great crowd had gathered,
He was loved by his friends and his neighbours and all,
To the grave went his corpse, by his friends it was followed,
And tears from our eyes like the dewdrops did fall.
The widow lamented the fate of her husband,
Broken-hearted she cried on her dear loved one's tomb:
'To the world there is left three poor little children,
Whose father had met with a coal-miner's doom.'

Heroes, British Heroes

We sing of our soldiers and sailors,
Brave deeds they have done o'er the foam,
But what of our lads who work down the mine?
Little of them may be known.
They start out for work in the morning
With many a joke and a laugh,
With never a thought of danger
As they go down the shaft.
    They are heroes, British heroes,
    Hearts that are brave and bold,
    Risking their lives for children and wives,
    Down in that blackened hole.
    Toiling for mere existence,
    Where the sun it never shines.
    Heaven protect our collier lads
    That work down in the mine.

With scarcely a moment's warning
A cry it rings through the air.
Like wildfire it spreads, the pit is on fire,
With four hundred souls working there.
Alas, there's no hope for those miners,
The rescuers no more can do.
They die just like true British heroes,
For home, wife and children too.
    They are heroes, British heroes,
    Hearts that are brave and bold,
    Risking their lives for children and wives,
    Down in that blackened hole.
    Toiling for mere existence,
    Where the sun it never shines.
    Heaven protect our collier lads
    That work down in the mine.

To the gallant deeds of those heroes
The limit no one can tell.
At Pretoria Pit and Whitehaven too,
Many a brave collier fell.
And when the peacemaking King Edward,
Who through God's mercy now lies at rest,
Awarded his medal of valour,
It was to heroes the bravest and best.
    They are heroes, British heroes,
    Hearts that are brave and bold,
    Risking their lives for children and wives,
    Down in that blackened hole.
    Toiling for mere existence,
    Where the sun it never shines.
    Heaven protect our collier lads
    That work down in the mine.

Collier Heroes of the Mine

Our sailors on the wave, our soldier lads so brave,
Against our foes they gallantly combine.
But down below the ground, there brave men can be found;
I mean our collier heroes of the mine.
    My song is of a boy, a widow's only joy;
    His father and his brothers three were dead;
    But the widow's heart was sad for her own dear collier lad
    Who went down in the mine to earn her daily bread.

    How bravely the collier toils, digging for coal,
    As day and night goes round.
    But ere the day has fled, how many may be dead.
    May God protect our colliers in the mine.

One day a dreadful roar just seemed outside the door.
The widow sobbed: 'God help my only son!'
As she joined the crowd outside, 'The pit's on fire!' they cried.
She went off to the coal-pit in a run.
    In tears they gathered round some bodies were on the ground.
    The cries of weeping females rent the air.
    Sons, fathers and brothers were found by their poor mothers
    Amongst the dead and dying lying there.

Down in the mine below, the widow tries to throw
Herself, for nothing of her son she hears.
But a cage full of men came to the top just then,
And a boy jumped out and dried away her tears.
    With heart no longer sad, she welcomes her dear lad,
    Although they cried for others' grief and woe.
    'Oh, may heaven help the brave, who have found a collier's grave,'
    They cried so bitterly as they homeward go.

    How bravely the collier toils, digging for coal,
    As day and night goes round.
    But ere the day has fled, how many may be dead.
    May God protect our colliers in the mine.

Rap Her te Bank

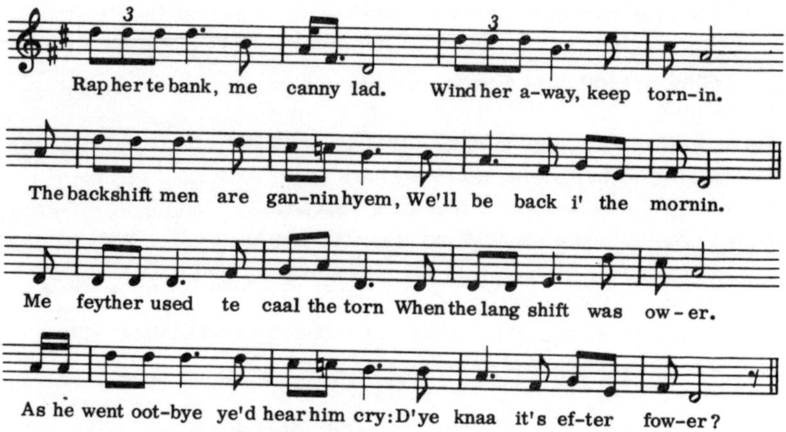

Rap her te bank, me canny lad:
Wind her away, keep tornin'.
The back-shift men are gannin' hyem.
We'll be back in the mornin'.

Me feyther used te caal the torn
When the lang shift wes ower.
Aal the way ootbye ye'll heor him cry:
'D'ye knaa it's eftor fower?'
Rap her te bank, me canny lad!
Wind her away, keep tornin'.
The back-shift men are gannin' hyem.
Aa'll be back in the mornin'.

And when that aaful day arrived,
The last shift for me feyther;
A faal o' stones an' brokken bones,
But still above the clatter he cried:
    Rap her te bank, me canny lad!
    Wind her reet slow – that's clivvor!
    This poor aad lad is tekken bad.
    Aa'll be back heor nivvor.

The Man Called Len

There was a man, his name was Len,
Got buried in t'rippin at quarter to ten.
They tied him up so good and well,
But who it was they couldn't tell.

Up the drift then they did go.
One end was Steve and the other was Joe.
They got to t'top and there stood Ben,
And Ben said to Len: Nay, it's not thee again?

Ay, it's me again, said Len to Ben,
I got buried in t'rippin at quarter to ten.

VI

Coalfield Conditions and the Struggle for a Better Life

VI

COALFIELD CONDITIONS AND THE STRUGGLE FOR A BETTER LIFE

The ballads of discontent are a considerable part of the body of mining songs. Some went no further than to bemoan the sorry conditions of the miner, his low wages, long hours, dear food at the tommy shops. Some were concerned with dreams of escape to the goldfields of California or New South Wales. But most of them speak of political action, where men are fighting to get something done.

In particular, events in the North-East of England and the resultant enthusiasm for solidarity that grew up so powerfully during the 1830s and '40s inspired a number of miners – and some miners' wives – to try their hand at songmaking. In some ways the 1844 Durham strike was the first modern-style stoppage, when the spirit and technique of trade unionism were beginning to show in a way we are familiar with today, and the balladry of this strike was to set a pattern for the bards of later struggles in the coalfields. Chartist agitation was loud in the air. The first national organization of pitmen had come into being – the Miners' Association of Great Britain and Ireland, formed at Wakefield in 1841. The new-discovered solidarity, the clearness of aim, the decent sense of responsibility in the face of oppression and provocation, show clearly in the ballads of the time. Curiously enough, the epic hero whose name cropped up again and again in the songs was not a pitman but a lawyer, the Chartist William Roberts, engaged as legal adviser by the Miners' Association, who championed the trade union cause in the law courts and was looked on by the miners as a true friend and a fine fighter.

The non-union miner, the blackleg, appears in the balladry of the last hundred years and more. In some cases, Cornish, Irish or Welsh labourers were imported as strikebreakers in English and Scottish coalfields. The miners might hunt the blacklegs over the countryside, strip them, and blacken their faces with candle-grease and coal-dust. One song tells of more dangerous forms of revenge in the darkness of the mine, with ropes stretched neck-high across the way, to catch the non-union putter by the throat and fling him backwards as he hurried with his tubs.

During the twentieth century, the making of strike songs seems to have declined; even the bitter dramatic struggles of the 1920s inspired few lasting songs, and the typical ballads of discontent in the 1930s were ephemeral parodies of popular song-hits, *Bye Bye Blackbird* in particular. Since Nationalization, a characteristic kind of song, sadly enough, has been the farewell to pits closed down by the Coal Board, with the consequent break-up of old mining communities, and the transfer of workers to new regions where they must modify their native dialect, learn to like a different brew of beer, and march under unfamiliar union banners.

The Pitman's Happy Times

When aw wes young, maw collier lads,
   Ne man cud happier be;
For wages was like sma' coals then,
   An' cheps cud raise a spree.
Wor pay-neet cam' wiv drink an' dance,
   Wor sweethearts luckt se fine,
An' lumps o' beef an' dads o' duff
   Wes there for folks te dine.
An' then we spent sic merry neets,
   For grum'lin' we had nyen;
But the times o' wor prosperity
   Will niver cum agyen.

Wor hooses then wes ower sma',
   Fir ivery neuk was chock;
Wor drawers wes fair mahogany,
   An' se wes chairs an' clock.
Wor feather beds an' powls se fine,
   Wes welcum te the seet;
A man worked harder i' the day,
   Wi' thinkin' o' the neet.
Spice hinnies on the girdle fixed;
   Maw tea had rum in't then;
But the times o' wor prosperity
   Will niver cum agyen.

Wor wives cud buy new shawls an' goons,
   An' niver heed the price;
The spyed-yece guineas went like smoke
   Te myek wor darlins nice.
The drapers used ne tickets then,
   The country gowks te coax;
They got thereckly what was ask'd,
   An' praised us collier folks.
The butcher meat was always best

When Kenton paid thor men;
But the days o' wor prosperity
Can niver cum agyen.

When aw got wed – gox, what a row!
The bindin' brass aw spent;
Aw bout new gloves an' ribbons, man,
For aw the folks aw kent.
At ivery yell hoose i' this toon,
We had a cocktail pot;
Wi' treatin' a' the company roond,
Maw kelter went like shot.
But, smash! we had a merry neet,
Though fights we had but ten;
Thor wes sic times for collier lads –
They'll niver cum agyen.

We didn't heed much lairnin' then,
We had ne time for skyul;
Pit laddies worked for spendin's syek,
An' nyen wes thowt a fyul.
Noo, ivery bairn can read an' write –
Extonishin' to me!
The varry dowpie on ma lap
Can tell his A B C.
Some folk gets reet and some gets wrang,
Biv lettin' buiks alyen;
But this aw'll sweer, ne times like mine
Can iver come agyen.

The Miners' Binding

It happened on March the twenty-third day,
When our binding came on, and we all went away,
And our miners of Backworth they all did agree,
And the number of them it was one hundred and three.
As I've told you before we agreed to a man,
To be true to each other and firmly to stand,
But we broke thro' our rules and began for to bind,
Since we're all brought to nothing on both Wear and Tyne.

Our miners and keelmen, and sailors also,
Since our trade has come down are all very low,
The like of this binding we never did see,
Since the coal-trade commenc'd by land and by sea,
To take off our money I'm sure 'twas not right,
We lament our misfortune by day and by night,
When we push'd round the glasses but now must decline,
Since we're all brought to nothing on both Wear and Tyne.

We thought that the year nineteen it was bad,
It was all the whole cry, there's no money to be had,
But it's nothing at all to this year twenty-two,
It's all the whole cry, alas! what shall we do,
The like of those times was ne'er seen I dare say,
Since the evening and morning became the first day,
The poor's been oppress'd this very long time,
Since we're all brought to nothing on both Wear and Tyne.

Tho' our binding money they've taken withal,
We may see better days yet in spite of them all,
To make themselves gentlemen I do protest,
They do not know what way the poor to oppress,
But since it is so we must do as we can,
We hope to see better days every man,
This year twenty-two now we always may mind,
That we're all brought to nothing on both Wear And Tyne.

We're all brought to hardships as plain you may see,
Which makes us lament now in every degree.
This binding we've had on our own native shore,
We ne'er will forget till our time is no more.
The way which now we are oppress'd every man,
It makes us down-hearted, do all that we can,
Our wages is low and trade is on the decline,
And we're all brought to nothing on both Wear and Tyne.

So all you brave miners wherever you be,
I hope there is no one offended at me,
Altho' I did venture to pen out these lines,
I was urged to it now by the change of the times.
My song for to finish I will add little more,
But may the Great Ruler of both sea and shore
Bring our island to flourish in a very short time,
And our trade it will rise then on both Wear and Tyne.

The First Drest Man of Seghill
(or: The Pitman's Reward for Betraying his Brethren)

Come, all ye miners far and near, and let us all unite, o,
In bands of love and unity, and stand out for our right, o.
Like Israel, these many years in bondage we have been, o,
And if we do not still stand out our truth will not be seen, o.
With my fal lal la, etc.

No doubt but there may be some men not bad for to deceive,
But to their ruin it will be, just like our mother Eve.
Our mother Eve was sore deceived by one that's called the Devil,
And by persuasion she was brought to do the thing that's evil.

Man a weak frail being is and easy to deceive,
And by a man called black J. R. was made for to believe –
It was on March the nineteenth day, eighteen hundred and thirty one,
A man from Earsdon Colliery his brethren did abscond,

And to the Seghill binding he did come with all his might,
For to deceive his brethren dear he thought he was but right.
But when he came to Seghill town the men were standing off.
He thought that he would then be bound and he would make a scoff.

As other men were standing off, he would not do the same.
That idle work would never do, he'd rather bear the shame.
Black J. R. made him believe that he was in no danger,
And to the office he might go, because he was a stranger.

He to the Seghill office went all with a bad intent;
But ere he got to Murton town he was made to lament.
He to the office went with speed, as hard as he could batter,
And other two young lads with him – we did give him a tatter.

Black J. R. made him believe that men would not molest.
He being acquainted with the man, he was by faith possessed.
We kept the fugleman in view, which made us all so keen,
For it's an oath by brethren all, no one they are to screen.

We went in chase of this bad man for to give him a scar –
By coming to our bond, you know, to serve the man J. R.,
And at the hour of two o'clock as I was sitting cobbling,
A rout there came unto our house, I heard the women gabbling.

Away I went with all my speed as hard as I could hie.
To see if I could catch the hares it was my will to try.
We gladly found him at the pit with J. R. and G. S(oth)ern,
But our hearts were fully bent not to stand there bothering.

But there were some upon the chase long ere I got there.
With running so I lost my breath, and I could run ne mair.
But I will tell his travels here as he went from the binding;
They stripped him there of part his clothes and left his skin refining.

Black J. R. was all the blame; he lost all but his lining,
But when he came to Hallowell his skin so bright was shining.
They left him nothing on to hide the good old man, the priest,
But there they put him on his hat; he was so finely dressed.

They set him off from there with speed to an ale-house by the way,
And there the Earsdon men did sit a-drinking on that day;
But what their minds I cannot tell when they did see him coming.
The priest he had within his hat and hard he then was running.

And all the way as he went home, by many was heard say
That persuaded he had been to his loss upon that day.
The Earsdon men they set him off from there to the Machine
That stands upon the allotment hill; he there himself did screen.

Then to the left hand road he took, the road that leads to Murton,
And there under a good whin bush the priest and him sat lurking.
The priest was almost starved there – I'm afraid he has got harm.
'If I was only into bed, I quickly shall be warm.

But unto Paradise I'll go where I will get no ill,
But nevermore will I go back to that place called Seghill.'
But remember you that come unto Seghill to bind –
You may think upon the man that we have treat so kind.

The Pitman's Complaint

O Lord hear the poor pitmen's cry
Look down on us with pitying eye;
With heavy bondage are opprest,
And all our families are distrest.

Thou heard the Israelites of old,
And led them to a blessed fold;
Deliver us from slavery
And set the Sons of Britain free.

In the dark pit where we are bound,
The iron hand of oppression's found;
Our labour's hard, our wages small,
Some days we work for nought at all.

As lions greedy of their prey,
They take our rights from us away;
To starvation we are driven,
Pale and wan we are ill thriven.

Our masters pinch us very sore,
We never felt such smart before,
They have us so completely bet,
Not one in fifty's out of debt.

The Indian slaves for Freedom groan,
We have a greater cause to moan,
You often pity slaves abroad,
But we have now a greater load.

Come, O ye rulers of our land,
Pray take our cause into your hand,
Then let us have fair Britain's law,
And save us from proud Pharaoh's paw.

Beneath the harrow we are crush'd,
Our blood lies mangled with the dust;
Regardless of our cries and groans,
They suck the marrow from our bones.

As cannibals they have eat our flesh,
Their bellies swell to great excess,
To quench their thirst have drunk our blood,
And left us wallowing in the mud.

Does not the trumpet sound reform,
And are we not free Britons born?
We want to have a jubilee,
The slavish pitmen now set free.

Our flesh pots now are stained with rust,
Our cup-boards now without a crust.
The tears run from the mothers' eyes,
They can not bear their children's cries.

Arise my brethren from the dust,
And in the Lord let's put our trust,
Then all our foes he will confound,
And in the sea proud Pharaoh dround.

The Pitmen's Stick

Ye collier lads, I pray attend
To what I'm going to say,
For I am very glad to hear,
That you have won the day.
Great many people did believe,
That you would get the worse,
Because they thought your masters
Were too strong in the purse.

But let their purse be e'er so long,
Or be they ever so wide,
I always said that you would win,
For the Lord is on your side;
For those who're in a righteous cause,
Have never cause to fret,
For when the Lord is on their side,
They never can be beat.

For many years we know it well,
You have been oppress'd,
You may thank the blessed Lord
That you have got redress'd.
I hope those lads will be thought on,
More than they were before,
For we've a right for to be glad,
They've started work once more.

For they are the main support,
Of this whole country side.
I'll sing their praise where'er I go,
Throughout this country wide.
The merchants in Newcastle town,
They were greatly distress'd,
For they thought the collier lads,
Would never be redress'd.

The landladies long faces made.
  Good right they had I know,
For many a pound those lads did spend
  When they had work below.
If they'd been much longer off work
  Our fate would have been bad;
For you know this to be true,
  Coals were not to be had.

Full seven weeks their stick did last,
  You know this to be so,
Before that any man did ride,
  For to work down below.
Great praise the pitmen all have got,
  Throughout this country round.
May God protect the collier lads,
  When working underground.

Their masters would have more feeling
  For their men I know,
If they had but one month to work,
  With their brave men below,
And be expos'd when down at work,
  To a blast of foul air;
Their prices they would ne'er abate,
  But sooner give them more.

Now their strike is over,
  And they've got their desire,
I hope we'll never more have cause,
  To cry out for want of fire;
And I hope they will ne'er have cause
  For to stick any more,
But live in comfort all their lives,
  As their friends have done before.

There never was so many men,
  Which every tongue can tell,
During the time of their strike,
  That e'er behav'd so well.
So let us drink a health to those
  Brave lads who hew the coal,
For I think they should be well paid
  For working in the hole.

The Blackleg Song

As traitors we shall know them now.
Their principles are dust,
For they have broke their taken vow,
And the rules of workers' trust.

Their action now has likened them
Unto the slaves of old
Who had to be content to work
While masters took the gold.

They sneak out quiet to their toil
At night, also at morn,
Their faces all dejected,
For they're looked on now with scorn.

We'll bid them all a last goodbye.
Their company, we can spare it,
For we never want the rotten dogs
To uphold trade union merit.

So now their names are on the scroll
Of dirty blackleg fame,
And we hope their innocent children
Will recall the day with shame.

The Pitmen's Union, or The Lads of Wear and Tyne

Now let the colliers' hearts be glad
While plenty round them shines,
And blest contentment flows along
The banks of Wear and Tyne.
    Still round our banners we will stand,
    In love and truth combine,
    And children yet unborn shall sing
    The lads of Wear and Tyne.

Brave Hepburn and our delegates
Like rays of virtue shine.
Their fame shall long be echoed round
The banks of Wear and Tyne.

On Bolden Fell our flags shall wave,
Like victory's wreaths entwine,
But peace shall be the motto still
With lads of Wear and Tyne.

We envy not the rich and great
Whose dazzling greatness shine,
While we, the hardy sons of toil
Can labour in the mine.

Our happy wives and children now
All former cares resign,
And sing with joyful mirth and glee
The lads of Wear and Tyne.

May he who rides upon the storm
Protect with care divine
From all dangers that surround
The lads of Wear and Tyne.

Here's a health unto the king,
Likewise the queen sublime,
Who gave the pitmen their applause
That dwell on Wear and Tyne.

Now to conclude and make an end,
May luck around them twine.
Oh, bless the happy collier lads
On both the Wear and Tyne.

The Oppressions of the Pitmen

Eighteen hundred and thirty-one, as I have heard them say,
The pitmen of the Tyne and Wear met on a certain day,
To consult about their wages, which makes me for to sing –
Expecting for to meet them upon Shadon's Hill.
    The oppressions of our country
    We cannot now endure;
    The viewers they get all the brass,
    And we are kept so poor.

The owners and the viewers are all a wicked crew,
They have been long contriving to punish me and you;
To punish me and you, as I have heard them say,
The viewers of the collieries caused men to run away.

Some have gone to America and Ireland also,
For to maintain their families to Scotland they go,
For money in our country is now so bad to get,
The viewers they must have it – and so we go to wreck.

We are meeting about our 'greements – mind what I do say,
And if we do not stand true, we'll have to rue the day:
Our viewers they'll be harder still, our prices to reduce,
And if we do not stand true our meetings are no use.

There was Hetton and Russell, and Renton also,
Some from Piddington and Lambton, and some from Shiney Row;
There were from Harrington and Fatfield, and Ouston that we ken,
They all came for to meet upon the Black Fell.

When they came unto the Fell to each other did say,
We must have a committee to settle all this fray;
The committee was agreed on, that you shall understand –
Down the Fell to the Mill-house the delegates did gan.

When we came to the Mill-house, H——n he did say,
There's two or three things in our bond we must have done away
There's lying off the three days before we can demand,
And sixteen hours for bits of lads – I think it's ower lang.

There's many a man binds himself, and does not know what about
And when he is out of employ, the house they turn him out;
The house they turn him out, as I have heard them say,
The best thing is now to strive to have them done away.

When we came unto the Hill, Tommy U——l for to hear,
He read some papers unto us, which was very queer;
He said the viewers had a club, and other things we ken;
It's all to keep the owners blind as well as the poor men.

There was Willy D——n, a man that we did ken,
Said that he would bind himself – he would not mind the men;
He being a treacherous rogue, besides a wildish rake,
They stripped the clothes all off his back, and sent him home naked.

Now our times are all out, and dutiful we have been,
Such union amongst the pitmen before was never seen;
Such union among the pitmen the world did never ken –
The oppression of viewers causes this amongst the men.

Come all you bold pitmen, since it has come so near,
If we do but stand true, we'll have no cause to fear,
The viewers they'll have to comply, for they will plainly see,
That there will be a stick, if they do not agree.

The Pitmen's Agreement

Come all you honest pitmen lads,
   And listen a short while;
It is concerning the agreement,
   Your time I will beguile.

A glorious victory you have gain'd
   Over your stubborn foes,
(Makes me sing these lines with glee)
   Which everybody knows.

Some of you have been badly used,
   Most shamefully used indeed;
For out of doors you have been turn'd
   And left you without bread.

Some of you have been turn'd out of doors,
   Which I do still maintain,
Complying not with the demands of those,
   Whose house you did retain.

During the whole of your manly strike,
   Your conduct have been brave –
Which deserves the country's thanks,
   And the nation's praise.

All those men that are not bound
   Cheer up your daring hearts –
Patience and Perseverance, my lads,
   Be still your nobler parts.

Should you any act of violence commit,
   A stigma on you will be;
The military force will interfere,
   Which you will plainly see.

All those men that have got bound,
   For their aid you must call,
To give you part of their earnings,
   For fear that you should fall.

After you have all agreed, my lads,
   (I hope you'll excuse my rhyme)
In a sacred bond join heart and hand,
   To support you 'nother time.

Now success to all the pitmen,
   Both on the Tyne and Wear;
May they continue to have their rights
   In each succeeding year.

Success attend their canny wives,
   And all their children dear,
Forgetting not the Delegates,
   Who had the cause to steer.

The Tools of Shotton Moor

A song I'll sing if you'll attend and pay attention to my ditty
To these few lines that I have penned, to waste my time it is a pity:
I sing of things in form of men, but they of men have not a feature,
For God did never man ordain, to undermine his fellow creature.

At Shotton Moor we have some tools, for men it is not fit to call them,
They think that everyone is fools, but mark and see what will befall them;
There's some of them been much esteemed, both far and near their name has sounded,
But little have they thought or dreamed how their character now is wounded.

There's one perhaps you know him well, a preacher was, perhaps you've heard him,
For gain his principle did sell; for it the Lord will sure reward him;
He has not kept the law so good, which we as well as him should go by:
Do to thy neighbour as thou would, that thou would wish for to be done by.

Then little Crooks has got a face for everyone that he does talk with.
He has proved it since he got a place, and a yard wand all for to walk with.
To act the tyrant he knows how, but he had better stop his scoffing,
Lest overboard we him should throw, then, dirty tool, what will come of him?

Then Willy Gape, a Cockfield man, he is a base and treacherous villain.
To hurt us he'll do all he can, but to do good he is never willing.
A dirty useless tool he is, he is always trying to deceive ye.
He should be drowned in the sea, he is not fit to live, believe me.

Then Wardle Tom's next in the clan, for favour he'll do ought that's dirty
To undermine his fellow man, but men like these can never hurt ye.
False to his neighbour he has been, now mark and see what will come on it:
He'll rob his master, 'twill be seen, give him the chance, depend upon it.

And ruff-faced Jack did act his part, as well as any that I've named,
To work he never had the heart, for worthlessness he has been famed,
He once was called a bonny clock, but like something that's out of season,
The pendulum's not worth a shock, it was so short and that's the reason.

Then plain-face Bates so bad did show, when to the magistrates they took him
He cried out: 'Stop! to work I'll go,' just as the clerk was going to book him;
'I cannot think to go to gaol, although resolved when I did enter,
But now I feel my strength to fail, for I have not the heart to venture.'

The Rechabite Man

Come sit you down and hear the news,
And screw your courage tight, man;
And I'll tell you of that rattlesnake,
Poor Bob the Rechabite man.

Now Bob he was a ploughboy
When ploughing was delight, man.
'Twas then he came to Biggesmain
And joined the Rechabites, man.

We thought he was a genius,
With his buckles shining bright, man.
His looks did show the hypocrite,
Poor Bob the Rechabite man.

Now when he came from Biggesmain
'Twas rather late at night, man.
This hypocrite he did come here
The horses for to keep, man.

Now when the horses they were nearly done,
He was often on his prey, man,
And because he was a hypocrite
He got in the rolleyway, man.

He stopped there a good few weeks,
But it was all in vain, man.
He had to take the pick, it did his trick,
Hewing at the Bankhead Crane, man.

He corved her in at the right renk.
No vantage could he make, man.
He told it to the putter lads
That he had got a sore head-ache, man.

He quickly put up his gear.
He buckled them so tight, man.
He thought it was a better plan
For to be in the Rechabites, man.

And then to bank poor Bobby came
And home he went with speed, man.
Bella wrung her hands and gave a scream
That Bobby he was dead, man.

She bathed his brows with vinegar,
And laid him into bed, man.
She lapped him up in the cradle twilt,
And bade him rest his head, man.

Then the Doctor he was sent for,
A chap of noted skill, man.
The neighbours they were all called in,
Because poor Bobby he was ill, man.

The Doctor he did soon appear;
The day was now far spent, man;
Took out his watch and felt his pulse.
He could not find out his complaint, man.

The neighbours stood amazed
And stagnated to the brain, man.
'Twas with the freight that Bobby got
Hewing at the Bankhead Crane, man.

She placed her hands upon her hips.
'I have gotten such a freight, man,
No more shall he hew down the pit
As long as there's a Rechabite man.'

He lay in bed for three long months.
The nights were very dark, man.
Now he has broke the Rechabites,
And gone to shifty work, man.

Away to shift work Bobby went
Without either fear or doubt, man.
He stopped there but a short time
Until he was wrought out, man.

The men they all felt for him
And pitied his case, man.
If it had not been for the union
He would had to left the place, man.

Now all the thanks that we have got,
I'm sure it is not our due, man,
He has broken up the Rechabites,
And joined the blackleg crew, man.

Now as you walk down Hebburn Town
Only keep your eyes before, man,
And you will see that bonny ornament,
It stands before the door, man.

Now twice a week she is cleaning it,
And makes it shine so bright, man.
It is the case upon my word,
Poor Bob the Rechabite man.

There's that bonny nursery before the door,
Of peeble stones and chalk, man.
Fine Bella she does cock her lugs,
All in the garden walk, man.

The people all as they pass by,
They look at her with spite, man.
They think upon the blackleg knave;
It's Bob the Rechabite man.

Now this is to conclude my song,
I have got nothing more to state, man.
We'll ne'er forget that hypocrite,
It's Bob the Rechabite man.

But that bonny day will come at last
When we will get our due, man.
We'll think upon that hypocrite
Likewise the blackleg crew, man.

Come All You Canny Pitmen

Come all you canny pitmen here, come listen to my song,
Of these few verses I have penned, you will not think them long.
Stick fast to your integrity, be sure and do not fail,
And we'll make the coal trade flourish, my lads, it shall go bail.

> Come, let us all be cautious and mind well what we do,
> And not forget the rights we got in the year 'thirty-two.

We in the year of 'thirty-two all for our rights did stand.
To be true to each other, we agreed to a man.
We did agree like Britons bold our rights to maintain.
We were true unto each other until the victory we did gain.

When we think of the days of old, about twenty years ago,
When plenty flowed among us, how it fills our heart with woe.
Eleven days a fortnight we did all get good employ,
But now we're glad of eight, it's little we can enjoy.

Our masters in the days of old upon both Wear and Tyne,
They gave us good encouragement, both money and drink to bind.
Let men and masters all agree as they have done before,
And then our trade shall rise again upon our native shore.

So all you canny pitmen lads, be all of one mind,
And then our trade will quickly rise upon both Wear and Tyne,
And we will get our rights again and all be in full glee,
And then we'll make the houses ring, and spend our money free.

Consider all the hardships now our pitmen have come through,
How sadly we have been pulled down in the year of 'thirty-two.
But cheer up your hearts, my pitmen lads, I hope you now will smile,
And we'll forget the weary time we've all had this long while.

We hope this year of 'thirty-six will put it to an end,
And free us of our hardships, and our trade will quickly mend.
Let men and masters all agree with harmony to combine,
And then we shall rejoice and sing upon both Wear and Tyne.

Now to conclude these lines, we will say little more,
And wish our trade would rise again as it has done before,
And men and masters would agree before that it be long,
For the Lord above, now, he does know that we're not in the wrong.

Mr Roberts the Pitmen's Friend

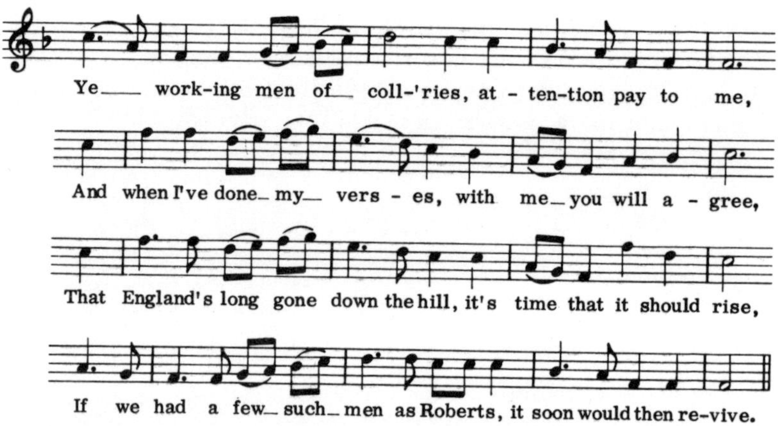

Ye working men of collieries, attention pay to me,
And when I've done my verses with me you will agree
That England's long gone down the hill, it's time that it should rise.
If we had a few such men as Roberts it soon would then revive.

The gentlemen of England have long kept down the poor
And bridged down all trades of every kind unto this very hour
For want of such men as Roberts to look about our rights and dues
For he's the man that will put them on their P's and Q's.

The pitmen a few years ago good wages they have got
Which made them blithe and happy and contented with their lot
But by degrees they've brought them down which caused them to say
They would rather be transported and sent to Botany Bay.

Their oppression was so great they could not it withstand,
Which made them join in heart and voice and likewise hand in hand,
And meetings rose in different parts their rights for to unfold
Then O'Connor thought of Roberts and his name unto them told.

Then to the North they brought him down their rights for to maintain
To stop the tyrants of their tricks, which no one could them blame
For down below they venture where the darkness is enrolled
And the treasure they do work for their masters for to hold.

Now as for the New Thornley men it's Roberts has stood true.
He has stood as great a contest as not many more could do.
Four days in Durham he did stand, their trial for to gain,
And for to release these prisoners out of the city gaol.

But their masters would not yield to them nor yet release the men,
So off to London Roberts went, the trial to extend.
His men he took along with him, the Queen's Bench for to see
And in less than half an hour his men in liberty were free.

Then the trial into chancery straightway did Roberts fling
And his men to view the city then along with him did bring.
Then to O'Connor he did go, took with him every man
And he treated them far better than the noblest in the land.

When back to Thornley they arrived the news for to unfold
The noblest sights that they beheld unto their men they told.
With hearts rejoiced they danced and sung, the music kept in tune
And these men chaired round Thornley streets all on that afternoon.

Now as for these poor Thornley men, no one should on them frown,
For sticking out for their rights which ought to be their own.
But we hope they will soon return again, the time is drawing nigh
And the north coal trade revive again, and away all tyranny fly.

Now the blacklegs I must give a rub before that I do end.
They will wish when we do get our rights they had been union men.
It is such as these blacklegs has caused all trades for to go down,
And caused thousands for to beg their bread, and the rich on them to frown.

Now here's to Mr Roberts, for he's the pitmen's friend.
He's worthy of his salary to make the tyrants bend.
Now may he live in unity and his men unto him cling,
For to take all tyranny from them, and out of them the sting.

A Dialogue between Peter Fearless and Dick Freeman

'Now, Dicky, was thou at the meetin today?
Aw looked, but aw saw nought o' thou by the way.
Aw hope that thou's not turnin cowardly noo,
Or yet thinkin o' gannin doon the coal-pit to hew.'

'No, Peter, aw have not been there, aw allow,
Or am aw yet thinkin o' gannin doon to hew;
But the fam'ly thou knaws, if aw didn't look out,
Would starve, wey, aw've had to put things up the spout.'

'Wey, man, such a meeting we've had, thou knaw.
Wey, the justice was there, an' he talked about law.
He said if the peace we kept, he would back us,
And if we did not, wey, the police should tak us.'

'Wey, Peter, tha knaws aw wad liked to been there,
What wi' redcoats an' bluecoats, 'twad be like a fair.
But noo aw will tell thou aw've got something to say:
There is things in wor bond that aw think out o' the way.'

'Things in wor bond, man; there's nought out o' reason.
Wey, man, aw's surprised, thou talks nowt but treason.
Now, where there's aught wrang, let's hear thou just say,
For aw think aw can prove there's nowt out o' the way.'

'Aw wad like for to hear, 'twad cheer up my poor heart,
If thou only could prove how the ten shilling smart
Is not far too much for the maisters to pay:
This is one o' the things aw think out o' the way.'

'Now, Dicky, aw'll let thou see what all this is for.
There is bad ventilation, an' wanting a door,
And a great deal more aw'll tell before aw do stop,
That the place where we hew, it very oft wants a prop.'

‘Wey, now aw do see, to prevent things like this,
That it’s to hinder the maisters from doin aught amiss,
But it’s to prevent the poor colliers from accidents heavy,
That we the ten shillings smart on our maisters do levy.’

‘Then the same thing applies to the rest that we ask.
That’s to hinder our maisters from wearin a mask.
They say that it’s accidents, when we know it’s neglect,
Then these things we ask for ourselves to protect.

‘Then let’s keep up our hearts, Dick, we are sure to win,
Our maisters aw knaw before long will give in,
Then let us stand fast, never stop by the way,
And victory we’ll shout when we win the day.’

The Miners' Grievances

A voice it has gone through the breadth of the land
That the Tyne, Wear and Tees they are all at a stand.
The miners have struck, for they've been oppressed,
And no more they will work till their wrongs are redressed.

For ten years or more we've been regular slaves,
Bound down by our masters, the greatest of knaves,
But no longer we'll bear it, for now we are free,
For right will beat might, that our masters will see.

No more will they get us to work by the fines,
For the splent and foul coal that we get in the mines,
And the word Separation won't stand in our bill,
For both roundy and small in the tub we will fill.

For their measures and steelyards, we'll do them away.
For by them they've robbed us for many a long day.
We will work by the weight now, for it is our will,
That a just beam and scale it do stand on the hill.

No more in the morning at one we will rise,
No more leave our homes till the sun meets our eyes.
The ten shilling smart we will have, it is true,
So that they may keep the place safe where we hew.

No more in their mines we will work without air,
For to have ventilation we'll make it our care.
Two shafts they may sink for the down and up cast,
To prevent the hydrogen from making a blast.

Then the rank of the putters is too far to run,
But I know that the sixty yard rank will be won;
Instead of twelve hours then the boys shall have ten,
That they may gain knowledge before they are men.

Now the guarantee clause, I think it will do:
That's the fifteen a week when we go down to hew.
Eight hours and no more we will stop in the mine,
And then come to bank without fear of a fine.

The five shillings a week I think is too small,
For the widow and family that lowses their all.
A coffin for those that is killed we do crave,
The five pounds to bring him with mence to his grave.

Now look at the dangers that we undergo,
While we work for our bread in the caverns below.
The rope it may break, the stone it may kill,
And the bursting of waters the pit it may fill.

The gas may ignite in the mine when we're under,
And cause subterranean lightning and thunder.
The choke damp it spares none that's left in the mine.
These things often happen in Tees, Wear and Tyne.

Now you see what we want, and I think you will say
That the miners are asking naught out of the way.
Then haste to the rescue, help the sons of the mine,
Cheer on the brave lads of the Tees, Wear and Tyne.

Martin Jude

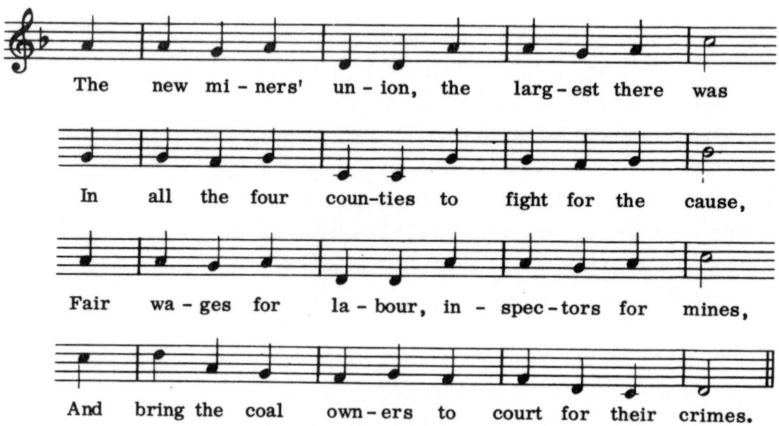

The new miners' union, the largest there was
In all the four counties, to fight for the cause,
Fair wages for labour, inspectors for mines,
And bring the coal-owners to court for their crimes.

A new fighting fund to go to the law.
Times they were changing, it was now forty-twa.
Jude was the leader, and to him they report.
Roberts the lawyer would battle in court.

Their grievances many by Thornley was shown.
In the northern coalfield resentment had grown.
Miners were robbed and the culprits went free,
Then added insult to their black tyranny.

A union's address to the owners was sent.
The answer received was a silent contempt.
Then marched the miners on March twenty-two.
The battle was old but the union was new.

There were twenty thousand bold mining men.
To Shadon's Hill they were marching again.
Once more united, they trudge through the snow.
Support for the union, come weal or come woe.

It was put to the vote – everyone for
Restriction of labour; the coalfield at war.
The first shot in the battle, the miners won't bind.
The voice of the dumb not heard by the blind.

More resolutions by Mitchell and Jude.
'Petition the Commons, it may do some good'.
You honourable men who in Westminster sit,
Take heed of the need of the poor in the pit.

From all the coalminers who toil in your land,
We humbly beg, sirs, please understand;
Send out your servants the mines to inspect.
From explosion, disaster the pitman protect.

Machines that pay pitmen who work in the dark
Rob them of wages for lack of your mark.
The law of today, to the owner it tells:
'You are the lawmaker, you can compel.'

The pitmen were scorned by the press of the day.
A journal that printed out Newcastle way,
A tool of the masters, for coal-owners' use,
On the bold Martin Jude poured scorn and abuse.

It made him a villain, as greedy as sin,
Because to the meetings no press was let in.
Held in Jude's ale-house; the point was well made,
But he's not blacklisted in the coalmining trade.

The delegate devils within the Three Tuns
On baccy and beer spent union funds.
Old Nick himself was vulgar and rude
As he gave a portrayal of bold Martin Jude.

It said that the law was ready to act,
To call in the militia when pitmen attack.
Jude gave his reply: To the masters you pander,
And twice you've been whipped for libel and slander.

When a miner speaks up to better his class,
You call him a fox, a wolf or an ass.
In the jungle of life, we fight to survive
Where parasites prosper and flunkies they thrive.

Jude soldiered on for mining reform.
Like every true captain, he fought many a storm,
But alas, did not live to see o'er the hill
That Parliament passed his Inspection Bill.

At Elswick churchyard he was laid to rest,
A champion of pitmen and one of the best.
Of all the praise paid, none could be finer
Than: 'Bold Martin Jude, a friend of the miner.'

The Ballad of Roberts

Come all ye bold miners wherever you be,
And give for a moment attention to me.
I'll sing you a song in the best way I can,
Of our brave lawyer Roberts, that praiseworthy man.

He's a dread to all tyrants wherever he goes;
They would sooner by half take a bull by the nose
Than meet him in contest or stand face to face,
For he baffles their wisdom and shows their disgrace.

At court when appearing he gives them surprise,
And causes the people to open their eyes.
He lays down the law in a manner so clear;
If he gets what is just, we have no more to fear.

He's so calm and so steady when stating a case,
And always stands up with a smile on his face;
For to meet his opponents he never is shy,
And when he gets justice, he soon puts them by.

He has been the best man that e'er came to our aid,
Since in the North Country among us he stayed;
For to conquer the tyrant he ever delights,
And give to us miners our own legal rights.

Before he came to us we were sore oppressed.
With unlawful reductions he had to contest.
By the hand of oppression we sore were abused,
Till we rose from our slumbers and boldly refused!

We could never get justice before he came down,
Which made the coal kings on the miners to frown.
Though oft we had laid down our case to them clear,
Our voices were useless, the kings would not hear.

They have long had their way, but we'll now have a turn,
For no longer beneath them we'll patiently mourn;
We'll teach them a lesson they'll long keep in mind,
And make them treat miners more gentle and kind.

We've gone forward like Britons and taken the field,
And a weapon called Truth we're determined to wield;
Till our rights we obtain, we will not stir an inch,
Then stand to your colours, brave men, and ne'er flinch.

A fair price for our labour is all that we crave,
And with God for our help we're determined to have;
It must not be said that the miners gave o'er
And lost the great fight in the year 'forty-four.

Whatever we do we must still keep the peace,
And abide by the law, never mind the disease.
Though we feel rather troubled about the blacklegs,
Let us once get to work and we'll give them the dregs.

So now, my bold lads, I must finish my song,
I hope as I've promised I've not kept you long.
Here's a health to bold Roberts, the miner's best friend,
And long may he live, all our rights to defend!

Come All You Colliers

Come all you colliers in this place and listen to my song,
And we'll show the viewers that they have done us poor miners wrong.
They have oppressed us very sore and that we truly know,
They've tied a knot they cannot loose, they've struck the fatal blow.
  Now is the time our wrongs to prove,
    While we in chorus join,
  Long may the union flourish still
    On Tees and Wear and Tyne.

The monthly bonds they offered us have forced us now to strike.
Six weeks now idle we have been, which owners do not like.
And if we go to work again, we vow to have our rights,
The six hundredweight tub we will regain, and no more sevens or eights.

Our price they lowered from time to time, which to our grief we know.
We'll ne'er forget their bounteous deeds wherever we may go.
They're beat this time, they all confess, our union's spread so far.
Against us they've combined for long; now sirs, we're on a par.

The blackleg men have proved untrue, and sought for rest and ease.
They've undermined their fellow-men, their masters for to please.
They hope for favours after this, and how they hew away!
But the time is coming fast, my lads, when they will rue the day.

There's one thing we have vowed to do, that is to keep the peace,
And not to trouble any foe, but friendship to increase.
Thus Roberts has exhorted us; by his advice we go.
He never led us yet to wrong, and never will, we know.

The Song of the 1844 Strike

Once compensation's bounteous hand
Lit up the gloomy mine,
And lustre lent our dingy band
In days of auld lang syne.

When right stood right and reason ruled,
Ere master did combine
To use quart pots to measure foul,
Bright days of auld lang syne.

With lightsome heart and upright head
I braved the dangerous mine,
But withered are all joys and dead
Sin' days of auld lang syne.

And now the noble-minded rank
Has grown handy with the fine,
We scarce dare send a tub to bank
Sin' days of auld lang syne.

And well reduction rule they've learned,
They one with all incline
To rob us of the half we've earned
Sin' days of auld lang syne.

But shall we thus in bondage bow,
In fettering bonds repine,
And bid one final last adieu
To days of auld lang syne?

No! Union's sun now sheds her rays
In lustre see her shine.
She'll light the path to happier days,
To days of auld lang syne.

Each man stand true, nor quit your post,
Ye sons of Wear and Tyne,
We'll yet enjoy the right we've lost
Sin' days of auld lang syne.

The Durham Strike
(Song to the Blacklegs and all those not in the Union)

Colliers! Where's your boasted glory?
Where's your manly virtue flown?
The renown in ancient story
While you store, as before,
Freedom or your bondage own!

England hath with Scotland flourished
When fair justice propped the state;
But when luxury was nourished,
Freedom fled, ruin spread,
Such may haply be your fate.

Selfish wishes soon are blasted.
By degrees they pine and fall.
Hateful while their branches lasted,
Now each hour prunes the power,
Tyrants vanquished one and all.

Viewers lusting after plunder
On impoverished vitals prey.
Honour's link they've torn asunder.
Virtue fails, vice prevails,
Certain signs of quick decay.

Think, oh think! Let heaven inspire you.
Banish avarice, pride and lust!
Let a manly courage fire you,
While unbound on Union's ground,
Stand you may, or fall you must!

The Coal-Owner and the Pitman's Wife

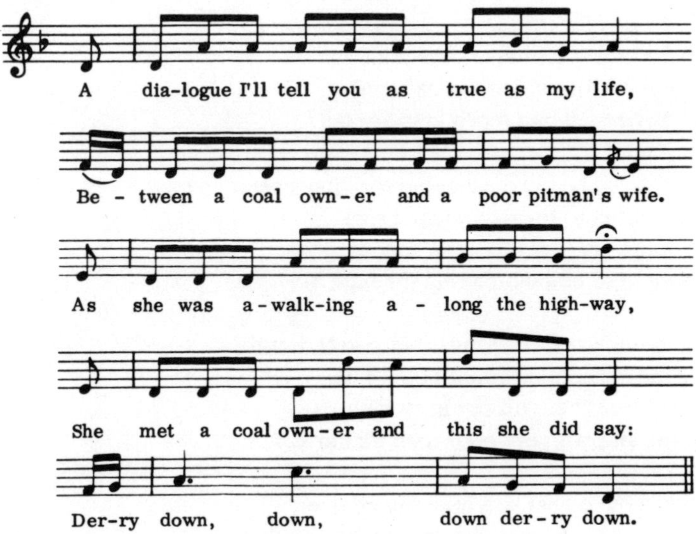

A dialogue I'll tell you as true as my life
Between a coal owner and a poor pitman's wife.
As she was a-travelling all on the high way
She met a coal owner and this she did say.
She met a coal owner and to him she said:
'Sir, to beg on you I am not afraid'
    Derry down, down, down derry down

'Then where do you come from?' the owner he cries.
'I come from hell,' the poor woman replies.
'If you come from hell, come tell me right plain
How you contrived to get out again?'

'Aye, the way I got out, the truth I will tell,
They are turning the poor folk all out of hell.
This is to make room for the rich wicked race
For there is a great number of them in that place.

‘And the number’s not known, sir, that is in that place;
And they chiefly consist of the rich wicked race
And the coal owners is the next in command
To arrive into hell as I understand.’

‘How know you the owners is next in command?’
‘How div I know, ye shall understand,
I hard the old devil say when I cam out
The coal owners all had receiv’d their rout.’

‘Then how does the old devil behave in that place?’
‘O sir, he is cruel to the rich wicked race.
He is far more crueller than you can suppose;
Aye, even a mad bull with a ring through his nose.’

‘Good woman,’ says he, ‘I must bid you fare well.
You give me a dismal account about hell.
If this be all true that you say unto me,
I’ll go home and with my poor men I’ll agree.’

‘If you be a coal owner, sir, take my advice:
Agree with your men and give them their full price.
For if you do not, I knaw very well,
You’ll be in great danger of going to hell.

For all ye coal owners great fortunes has made
By those jovial men that works in the coal trade.
Now, how can ye think for to prosper or thrive
For wanting to starve your poor workman alive?

So all ye gay gentlemen that’s got riches in store,
Take my advice and be good to the poor,
And if you do this all things will gan well,
Perhaps it will save ye from gannin to hell.’

So now the poor pitman may join heart and hand,
For when they're off work, all trade's at a stand,
Yon town of Newcastle, all cry out amain:
'Oh, gin the pits were at work once again!'

It's now to conclude, little more I've to say.
I was turned out of my house on the thirteenth of May,
But it's now to conclude and I'll finish my song;
I hope you'll relieve me and let me carry on.

The 1844 Strike

Come all ye good people that live in this village
And listen awhile to a tale that is true
Concerning the miners that all are laid idle
And swear so to be, till they get what is due.
For what do we ask but our rights in all reason?
To be paid for our labour is all we request;
And we're certain to get it if firmly uniting
We show we are men when we're put to the test.

They've lowered our wages almost to starvation
By enlarging the tubs and decreasing the pay;
We are worse used than other poor men in the nation,
For the schemes of our masters oppress night and day,
For thirty long years they have battled against us,
Till now we're united our rights to maintain:
And if we steer clear of the law, there's no danger,
Thus Roberts has told us again and again.

Ten weeks we've been idle and got no agreement,
And twenty weeks longer we'll stand yet and more
Tho' we're friendless and poverty starves all around us
Tho' we're houseless, deserted and turned to the door.
Like pilgrims forsaken the country we wander,
For a scanty subsistence we miserably go,
Yet we never repine, for we'll suffer a season
To conquer the tyrant that causes our woe.

They have promised to give better pay for our labour,
If the union we leave and friend Roberts desert,
But we know to our cost that they oft have deceived us,
So we will not regard the advice they impart.
We know our just rights and we vow to maintain them,
And the blow shall fall harmless they aim at our head,
With evils we'll grapple while evils are in them,
Or from principle perish as martyrs have bled!

So now to conclude the true things I've been singing,
Here's a health to brave Roberts wherever he goes:
He's a friend to the miner, a foe to oppression,
A Lion in contest, a Lamb in repose.
Then cheer up, my lads, for the day is fast coming,
When Victory will crown us and honour our cause:
Let your hearts and your voices in chorus responding,
Join the world that admires and repeats the applause.

The Pitmen's Union (I)

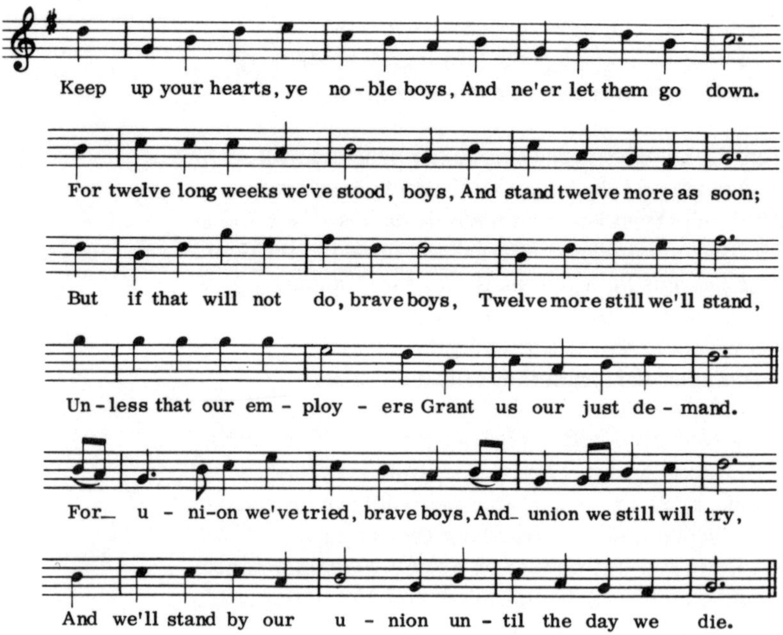

Keep up your hearts, ye noble boys,
And ne'er let them go down.
For twelve long weeks we've stood, boys,
And stand twelve more as soon;
But if that will not do, brave boys,
Twelve more still we'll stand,
Unless that our employers
Grant us our just demand.

> For union we've tried, brave boys,
> And union we still will try,
> And we'll stand by our union
> Until the day we die.

Roberts is our friend, brave boys,
To us he has proved true;
He's defended our just cause, boys,
In courts of Justice too.
He is a man of honour
The country well doth know –
He conquereth the viewers
Wherever he doth go.

There is the noble Marquis
Who sent for all his men,
To see if he could conquer –
But it was all in vain.
He tried every strategem
That was within his power,
And then he sorely threatened them
To turn them to the door.

He sent for them a second time,
But it was all in vain,
The men would not resume their work
Till they their rights did gain.
And when that he'd done all he could,
And threatened them full sore,
He sent the police and overmen
And turned them to the door!

The Pitmen's Union (II)

To our complaints give ear
While we our case relate.
To you it must appear
That our distress is great.
        To our petitions now attend
        And show that you're the pitmen's friend.

In Union we are joined
Up for our rights to stand.
Our masters are combined
To use the tyrant's hand,
        They long to see our Union end
        Which shows they're not the pitmen's friend.

Our Union gives offence,
So they its end would see.
All men of commonsense
Know what our state would be.
        Help us our Union to defend
        And show that you're the pitmen's friend.

You all must understand
That we indeed are poor.
Many by wicked hands
Are turned to the door.
        Who, who the helpless will defend
        And show themselves the pitmen's friend.

Long long may Mr Roberts live;
Long may our Union last.
May God our souls receive
When all our toils are past.
        Oh, may we all to Heaven ascend
        And reign with Christ, the pitmen's friend.

Pitmen's Grievances

Come all kindhearted Christians and listen to my song.
Such times in Durham ne'er were known and yet to last so long.
Our wives and children are turned out and camping out of doors,
Which causes us to wander and your charity implore.

'Twas on the 5th of April, that was the very day,
Our leaders had appointed our tools to bring away.
Until our wages are advanced we must work no more;
Which causes us to wander and your charity implore.

Our masters are hard hearted, our wages they'll not rise.
They will not hear us speak a word, our wants for to appease,
Unless our Union we will break and Roberts own no more,
Which causes us to wander and your charity implore.

Ye blacklegs of Wingate, I would have ye mind your ways.
You follow Christ for fishes, and then for gold betray.
Ye have sold us all for silver, and what can ye do more;
Which causes us to wander and your charity implore.

Here's a health to Mr Roberts, and long may he reign.
Likewise all other gentlemen who do our rights maintain.
Before our Union we will break and Roberts own no more,
We're resolved for to wander upon some distant shore.

The Colliery Union

Come all ye noble collier lads, wherever you belong.
I pray you give attention and you shall hear my song.
'Tis concerning of our Union Lads, for they have proved so true,
They have stood fast, man to man, we must give them their due.
So stick unto your Union, and mind what Roberts say,
If you be guided by his word, you'll surely win the day.

Little did the masters think that you would stand so fast.
They thought that hunger it would bite, you would give up at last.
But like the widow's crouse of meal that never did run out,
The Lord did send them fresh supplies that served them round about.

The masters they devised a plan their Union for to break.
It only made the colliery lads the firmer for to stick.
For when they thought upon the time that they'd been bit before,
Before that they would go to work they'd beg from door to door.

Then for to get the pits to work they have tried every plan.
Both Scotch and Irish they have brought, and every countryman.
But all the coals that they have got have cost them double pay.
Cheer up your heart, you colliery lads, you're sure to win the day.

Success to your commander, and Roberts is his name,
Since he has proved so loyal we'll spread about his fame.
Cheer up your hearts, ye colliery lads, he'll not leave you alone,
After he has eat the meat he will give them the bone.

Let's not forget young Beesley, a man of wit possessed.
He's gained the country's favour, for he has stood the test;
And let your day be e'er so dull, you'll see the rising sun,
For they will gain the victory without either sword or gun
So stick unto your Union and mind what Roberts say,
If you be guided by his word, you'll surely win the day.

The Blackleg Miners

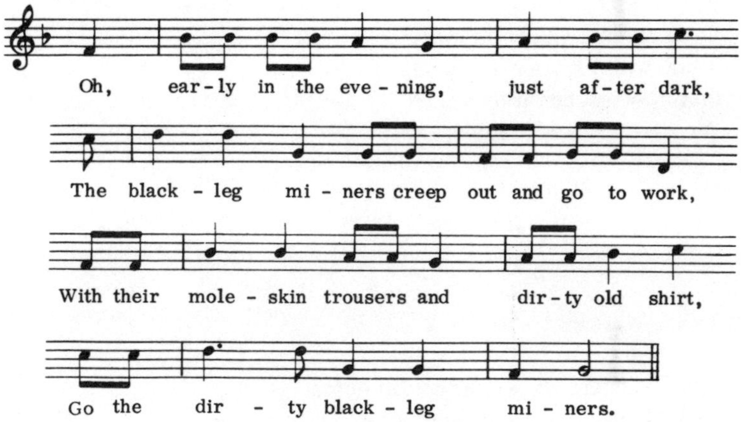

Oh, early in the evening, just after dark,
The blackleg miners creep out and go to work,
With their moleskin trousers and dirty old shirt
        Go the dirty blackleg miners.

They take their picks and down they go,
To dig out the coal that's lying down below,
And there isn't a woman in this town row
        Will look at a blackleg miner.

Oh, Delaval is a terrible place,
They rub wet clay in the blackleg's face,
And round the pit-heaps they run a foot-race
        With the dirty blackleg miners.

Oh, don't go near the Seghill mine
For across the mainway they hang a line
To catch the throat and break the spine
        Of the dirty blackleg miners.

They'll take your tools and your duds as well,
And throw them down in the pit of hell,
It's down you go and fare you well,
You dirty blackleg miners.

So join the union while you may,
Don't you wait till your dying day
For that may not be far away,
You dirty blackleg miners.

The Truck Masters

You truck masters all, pay attention to me,
Especially Gartsherrie and Summerlee;
And as I am sitting and thinking great long,
If you will listen to me I'll sing you a song.
As for some of our masters they are great men,
And one of them a ruler of our native land,
I wonder what's the reason he does not stop,
The infamous system of a truck shop.

The reason is obvious and plain to be seen,
The profits are great altho' they are mean, –
They rob the poor man in every stage,
And they very soon make him a small enough wage.
There is one in Drumpeller that place o' great fame,
But to sing about it, could I think shame,
For a collier there told me that works under ground,
That they subtract from him five shillings a pound.

There is one in Dundyvan as I've heard tell,
At charging a price none can them excel;
For butter and cheese is a penny a pound,
Above all other shops that you'll find in the town.
As for their meal and their bacon it's very good,
And fitting enough for a collier's food;
Their tea and their sugar is fair I declare,
But I sing no more of their grocerie-wares.

If a shilling or two you chance to take away,
They'll cross your book on the next cashing day;
They tell you quite freely you need not come here,
Such usage as this is hard I declare.
As Miller and Eadie sat drinking one day,
It's Eadie to Miller was heard for to say –
Drink up Mr Miller and send the glass round,
For the boys they'll pay for it that works under ground.

As for great Dixon he is a worthy man,
He keeps no truck shops to oppress his poor men;
When their money is earned he pays them cash down,
And I wish such a system was over this town.
So now you truck masters I bid you farewell,
Since I am not fitting your praises to tell;
But the next time I sing perhaps I'll say more,
I will tell you the grievances of every store.

The Collier Lad's Lament

In taking of my walks on a cold winter's day,
As through the collier's country I wended my way,
I overheard a collier lad, most bitterly he cried,
Oh, how I rue the day that my poor father died.

My father he worked in those pits here hard by,
When he did get good wages we had no need to cry,
But now the case is altered, the want of bread we know,
And soon unto the workhouse we shall be forced to go.

Myself and my poor brother in the morning we did go,
To work upon the coal-pit's bank all in the frost and snow,
The little that we both do earn is needless for to tell,
'Twill scarcely serve the one of us, the masters pay so well.

When my father he was living no tommy-shops were there,
He did receive good wages, and all things went on fair,
And when on Saturday he come home, he to my mother said,
Come let us go up into the town and buy our children bread.

But now the masters have the power to do just as they please,
And the collier he must labour hard, whilst they sit at home at ease,
And when their labour it is done the master thus will say:
You will not get your full amount of wages paid today.

Five and sixpence for a good day's work it was a collier's due,
But now he thinks himself well off if he gets more than two,
And if he grumbles at the price, the master thus will say:
To the workhouse with your children, and there get better pay.

The masters reap the benefit of all the colliers do,
By forcing from them much more than what's their due.
He does not think how they can live, and little does he care,
As long as all his work is done, they may live on the air.

To the tommy-shop they are forced to go for all that they do earn.
They are forced to take their wages out in bread and cheese and meal,
And when on Saturday they do go their wages to get paid,
The master says: The tommy do not forget today.

If the Queen and all the Ministers they all were for to come,
To live as these poor colliers do, and work underground,
And undergo the hardships and dangers of the fire,
I think they'd make the masters pay them better for their hire.

If Johnny Russell he was here and worked upon the bank,
And Albert he was doggy, 'cause he's of higher rank,
I think one week would settle them, and cause them thus to say:
Let these poor colliers have their rights, and give them better pay.

But when I do grow up a man if they don't better pay,
I'll go and be a soldier for thirteen pence a day
Before I'll work in those dark pits, and others for to share,
The benefit of what I earn in tommy-shops and beer.

But if any one's offended at what I now have said,
I only ask to give me work that I may earn my bread,
And pay us well when we have done, and then you soon shall see,
We do not sit at home and pine in want and misery.

Callerforney

Molly: Oh hinny, Geordie, canny man,
Thou kens aw likes thou dearly!
For thee aw turned off Baggy Crooks,
An' used Tim Targit queerly;
Billy Benson coaxed me sair to wed,
But man, aw cuddent spurn thee!
Oh hinny, canst thou think o' this,
An' gan te Callerforney!
Like honey blobs my heart'll burst
If thou gans te Callerforney.

Geordie: Hoots, Mally, haud yor whining gob,
Maw mind's myed up for sartin.
Maw picks and spyeds is i' my kist,
The morn aw's sure be startin'.
Aw'll seun be hikin' on the sea,
An' fleein' round Cape Horney;
Aw kens the seam to hew for gold,
When aw gets te Callerforney.
Oh, Callerforney, bonny Callerforney,
The very clairts upon the street
Is gold in Callerforney.

Molly: Thou's mevies rue, maw collier lad,
When in the waves thou's sprawlin',
When crocidiles and unicorns
Is at thaw hoggers haulin'.
Thou's not hae luck like Joney, man,
In some whale's guts to turn thee;
Thou'll lang to be wi' me at hyem,
An' far frae Callerforney.
Oh, Callerforney, shem on Callerforney!
Bob Stackers sweers thor's nowt but fules
Wad gan te Callerforney.

Geordie: Thou's wrang, aw tell thee, Mally lass,
    Just read the papers, hinny.
The place is verra like the Mint,
    Another Coast o' Guinea!
Though mind thee, yence aw heerd it telled,
    The cannibals wad burn ye,
And make gold ointment o' yor byens
    When ye get te Callerforney.
Oh, Callerforney, whei noo, Callerforney,
Hoots, Mally, aw can tresh them a',
    Aw'll conquer Callerforney!

Molly: Consider, Geordie, aw's thee wife,
    Aw diven't gan contrary.
If thou mun gan, thou's tyek the lass
    Thou ca's thaw bonny Mary!
But weel aw kens afore thou gans,
    Thou's trim'lin at the journey;
Sea-sarpints tee may cowp the boat,
    Then where's thaw Callerforney?
Oh, Callerforney, 'ticin' Callerforney!
Aw wish that folks wes not se poor
    To want thee, Callerforney!

Geordie: Cheer up, maw duck! thou'll gan wi' me,
    Aw niver heeds the danger!
Poor collier lads works hard for nowt
    An' still te death's ne stranger.
Like Whittin'ton, aw hears the bells
    That says, 'Come on your journey!'
Goold's better far than howkin' coals –
    Oh, dear, this Callerforney!
Oh, Callerforney, we're comin', Callerforney!
Farewell to splint, choke damp an' blast!
    Hurrah, for Callerforney!

Asstrilly (or: The Pitman's Farewell)

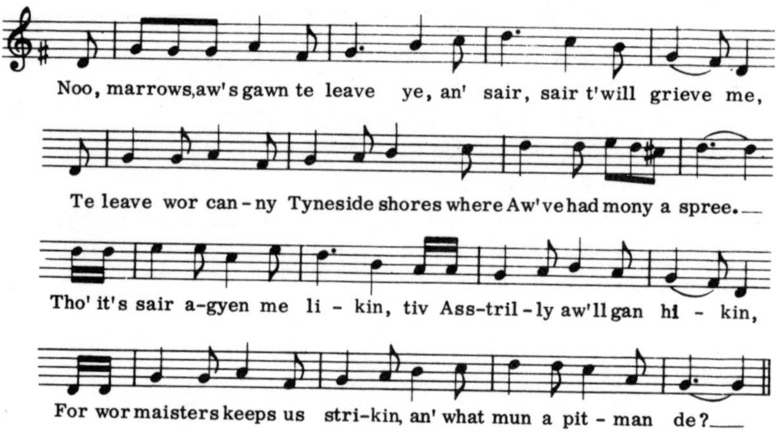

Noo, marrows, aw's gawn te leave ye, an' sair, sair 'twill grieve me,
Te leave wor canny Tyneside shores where aw've had mony a spree.
Tho' it's sair agyen me likin, tiv Asstrilly aw'll gan hikin,
For wor maisters keeps us strikin, an' what mun a pitman de?

Aw mind the time when collier lads cud work for goold at hyem, man,
Dash, aw mind the time when collier lads cud spend a pund each pay;
But noo the times they're queer, man, we've nowther sangs nor cheer,
man,
When we cannit raise wor beer, man, it's time te gan away.

Greet men may de a vast, man, but wor fine times tho're past, man.
Gosh, aw waddent leave wor canny toon, but aw's forced te gan away;
So aw'll myek ne mair emotion, but cross the salt sea ocean,
Where aw've got a kind o' notion when aw howk aw'll get gud pay.

Aw'll bid farewell te pit wark, an' howk for lumps o' goold, man.
Goshcab, aw'll suen be rich, aw've varry little fear;
So aw'll bid farewell te mammy, an' maw sweetheart o' the Lammy,
It's weel knawn aw's ne hammy – so tiv Asstrilly, lads, aw'll steer.

Oh, fare ye weel, ye happy scenes, where youthful days aw've spent,
man!
Fare ye weel, for better times 'boot here thor'll niver be.
So aw munnet be a gowk, man, but for goold aw'll gan an' howk, man,
Though maw boiley aw may bowk,* man, aw'll seun skim o'er the sea.

* To bowk one's boiley = to vomit

Bang the Bleazer, Ben

Bang the bleazer, Ben, bang the bleazer, Ben,
Cry: Fear ye not the masters, join the union men!

The year was eighteen forty-four, the big strike had been lost.
Trudged the miners back to work to reckon up the cost
Of twenty weeks' privation, not one concession won.
The men they were ten times worse off than when the strike began.
Bang the bleazer, Ben, bang the bleazer, Ben,
Cry: Fear ye not the masters, join the union men!

The union tried to rally, but meetings they were few.
There was no work for union men if e'er the masters knew.
To stir the old emotions on the Wear and the Tyne,
The brave bold old Ben Embleton kept banging out his rhyme.
Bang the bleazer, Ben, bang the bleazer, Ben,
Cry: Fear ye not the masters, join the union men!

The loyal lodge of Cramlington did union pits recruit.
'Rally round the banner, boys' played Ramsay on his flute.
One volunteer was Barrington, I'm sorry for to say.
The union and its leaders did ofttimes rue that day.
Bang the bleazer, Ben, bang the bleazer, Ben,
Cry: Fear ye not the masters, join the union men!

The pitmen pledged to keep the peace in eighteen forty-four
Even when the candymen turned them from their door.
But when the boys of Barrington did strike in forty-nine,
We heard the voice of violence cry for vengeance on the Tyne.
Bang the bleazer, Ben, bang the bleazer, Ben,
Cry: Fear ye not the masters, join the union men!

Resolving not to keep the peace, the peace of forty-four,
If strangers came to Barrington, the pit would work no more.
They said they'd mark the candymen, the flame of hate they lit,
With a warning to the blacklegs: You'll be buried in the pit!
Bang the bleazer, Ben, bang the bleazer, Ben,
Cry: Fear ye not the masters, join the union men!

With pickshafts at the ready they made their picket line.
Prepared they were for battle and for bloodshed on the Tyne.
The noisy boys of Barrington, vengeance was their choice,
And they made the village rattle with the anger of their voice.
Bang the bleazer, Ben, bang the bleazer, Ben,
Cry: Fear ye not the masters, join the union men!

One night the bold militia came and caught them in their beds.
Those strikers not arrested in disarray they fled.
At a single stroke the strike was broke, for Barrington no joys
In the tyranny of her masters or the violence of her boys.
Bang the bleazer, Ben, bang the bleazer, Ben,
Cry: Fear ye not the masters, join the union men!

The outrage seen at Barrington was seldom seen before.
They said that moderation lost the strike of forty-four.
But the brave bold boys who made the noise were called the foolish ones
When with pickshafts at the ready they took on the swords and guns.
Bang the bleazer, Ben, bang the bleazer, Ben,
Cry: Fear ye not the masters, join the union men!

The years gan by; we march on, aye, by eighteen fifty-two
Union pits were hard to find, although there were a few.
Later still, one stood alone; 'twas loyal Delaval.
The hardship of her union men was more than tongue can tell.
Bang the bleazer, Ben, bang the bleazer, Ben.
Cry: Fear ye not the masters, join the union men!

The Colliers' Eight Hour Day

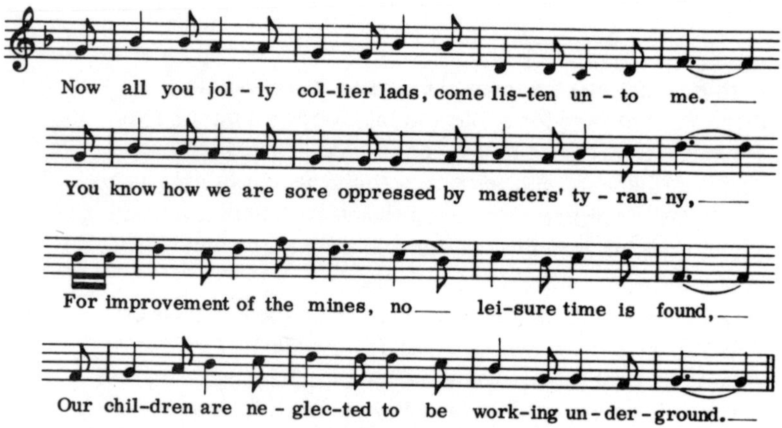

Now all you jolly collier lads, come listen unto me.
You know how we are sore oppressed by masters' tyranny.
For improvement of the mines, no leisure time is found,
And our children are neglected to be working underground.

> MacDonald is the colliers' friend, by us he true is found.
> He thinks eight hours is plenty for to work beneath the ground.

Our masters are tyrannical and that they must confess.
They overtax their workmen and do them sore oppress.
No other occupation so dangerous can be found.
We cannot call our lives our own while working underground.

The sailor he does plough the main and perils does go through,
But he sees the danger coming, which a collier cannot do.
With fallen roofs and firedamp the records can be found
How hundreds yearly lost their lives by working underground.

Down in the bowels of the earth our livelihood we gain,
Our wives and little families we toil for to maintain.
With five shillings a day some people say we'll never lack a pound,
But colliers cannot work full time, you know, beneath the ground.

MacDonald is the colliers' friend, by us he true is found.
He thinks eight hours is plenty for to work beneath the ground.

The most of trades and callings wherever that we go
Have gained the short-time movement, you very well do know,
And why should we poor colliers behind the rest be found?
We think eight hours is quite enough to work beneath the ground

Dalry, Kilbirnie, Johnstone, Irvine, Kilmarnock and Ayr,
Glasgow, Coatbridge, Airdrie and the whole of Lanarkshire,
Falkirk, Fife and Alloa and the Lothian men all say
That colliers should like other men but work eight hours a day.

Some useful books we will procure for to improve our mind,
Or health and recreation among the fields we'll find.
The leisure hours we have to spare will not be thrown away,
And we'll pray for brave MacDonald that has gained eight hours a day.

MacDonald is the colliers' friend, by us he true is found.
He thinks eight hours is plenty for to work beneath the ground.

The Miners' Cry

Did you hear the grand news that has passed in last session?
We are gaun to get justice like other trades noo.
The oppression we met with is past comprehension,
But a bright day is come, that appears to our view.
We have got good ventilation throughout the whole nation,
A great reformation is wrought on us noo;
Also the young of this generation will get education,
And that is something new.

> Cheers for McDonald, the Great Agitator,
> Who has fought for our rights against mighty men.
> The Buccleuchs and the Merrys may end their hurry-burrys,
> For Kinnaird and McDonald will beat them again.

Push on noo, Sandy, Dalglish will be handy.
You're sure of assistance from Pope Hennessy,
Shaftesbury and others, who have acted like brothers,
To be beat by the big one we never shall be.
For the cause that you have is to trample oppression,
And like our brave Wallace to set Scotia free.
The tyrants and their iron grasp of aggression,
They thought we should have no liberty.

Now pay attention, and listen with reason,
Take this to consideration, you might M.P.s;
If you are afraid for a foreign invasion,
Grant justice, we'll fecht them whenever ye please.
The brave hardy miners of Erin and Scotia,
We will take McDonald for captain and guide.
We'll put shouther to shouther and march with our thousands,
And there's no ane amongst us but would dee by his side.

But the struggle is not past yet, we'll hae to petition,
And Sandy, it's hoped you will gang back again
To show some mair of the miners' condition;
For short hours of labour you must try them again.
And though you do meet with stern opposition,
You met it before and you did get it;
That worldly tyrant, the Derby aspirant,
We'll chase him and beat him if union were in.

But we hope that our strikes and our struggles are ended.
We'll take social reform for our motto and guide
When our brave Kinnaird has the laws amended,
And it's hoped every man will stand by his side.
Now let us rejoice, there's a bright day before us,
The dark cloud is dispersed that hung in our view.
Let our thousands and thousands join in the chorus,
And every brave miner be loyal and true.

The South Medomsley Strike

If you're inclined to hear a song, aw'll sing a verse or two,
An' when aw's deun, you're goin to see that ivory word is true.
The miners o' South Medomsley they nivor will forget,
Fisick an' his tyranny, an' how they had been tret.
For in the midst of danger, these hardy sons did toil,
For to earn their daily bread so far beneath the soil.
To make an honest livin each miner did contrive,
But ye shall hear how they were served in hiteen hiety five.

The miners at South Medomsley, they're gan to make some stew,
They're gan to boil fat Postick an' his dorty candy crew.
The Maisters should have nowt but soup as long as they're alive,
In memory of their dorty trick in hiteen hiety five.

Below the county average then the men was ten per cent,
Yet Fisick, the unfeelin cur, he couldn't rest content.
A ten per cent reduction fre the men he did demand,
But such a strong request as this, the miners could not stand.
The notices was all sarved oot, an' when they had expired,
All the gear was brought to bank, the final shot was fired.
To hort his honest workin men, this low-lived man did strive;
He'll often rue for what he did in hiteen hiety five.

Fisick was detarmined still more tyranny to show.
For to get some candymen he wandered to an' fro.
He made his way to Consett, an' he saw Postick the bum.
He knew he liked such dorty jobs, an' he was sure to come.
Fisick telled him what to de, where to gan an' when.
So at the time appointed, Postick landed wiv his men.
Wi' polises an' candymen, the place was all alive,
All through the strike that Fisick caused in hiteen hiety five.

Commander Postick gav the word, they started wi' their wark,
But they were deun at five o'clock, they dursen't stop till dark.
An' when they had deun all they could, an' finished for the day,
The bobbies guarded Postick an' his dorty dogs away.
Fisick was a tyrant, an' the owners was the same.
For the torn oot' o the strike, they were the men to blame.
Neither them nor Postick need expect they'll ivor thrive,
For what they did to Dipton men in hiteen hiety five.

The miners at South Medomsley, they're gan to make some stew,
They're gan to boil fat Postick an' his dorty candy crew.
The maisters should have nowt but soup as long as they're alive,
In memory of their dorty trick in hiteen hiety five.

The Oakey Strike Evictions

It wes in November, an' aw nivor will forget,
The polises an' the candymen at Oakey's hooses met.
Johnny the bellman he was there, squintin roond aboot,
An' he placed three men at ivory hoose to torn the pitmen oot.
    Oh, what wid aw dee,
    If aw'd the poower mesel?
    Aw wid hang the twenty candymen
    An' Johnny that carries the bell.

There they went fre hoose to hoose to put things on the road,
But mind, they didn't hort thorsels wi' liftin heavy loads.
Some wid carry the poker oot, the fender or the rake,
If they lifted two at once, it was a great mistake.

Some o' these dandy candymen wes dressed up like a clown,
Some had hats wivoot a flipe, an' some wivoot a crown.
Some had ne laps upon their coats, but there was one chap warse –
Ivory time he had to stoop, it was a laughable farce.

There wes one chap had ne sleeves nor bottons upon his coat,
Another had a bairn's hippin lapped aroond his throat,
One chap wore a pair o' breeks that belang tiv a boy,
One leg wes a sort o' tweed, the tuthor wes cordyroy.

Next there comes the maisters, aw think they should think shem,
Deprivin wives an families of a comfortable hyem.
But when they shift fre where they live, aw hope they'll gan to hell,
Alang wi' the twenty candymen an' Johnny that carries the bell.
Oh, what wid aw dee,
If aw'd the poower mesel?
Aw wid hang the twenty candymen
An' Johnny that carries the bell.

candyman = rag-and-bone man, here a layabout
bellman = bailiff
hippin = baby's diaper

The Miners' Song of Freedom

Onward, ye brave miners, forward as to war!
Battles are not fought without an honest scar.
Tyrants must be punished, blacklegs made to run,
The future's bright before us, if firmly we press on.

> Forward be our watchword, let us all unite,
> And bravely fight for Whitehouse, for Union and for Right.

Onward, ye brave miners, but cast a look around.
We've left a dreary desert, and refuge we have found.
We've toiled hard for our masters, and offered up our lives.
They studied not our comfort, our children's or our wives'.

Onward, ye brave miners, and dare to do the right.
Oppression stands behind us to goad us to the fight.
Be men, and do not falter, for wrong cannot endure
The darts of honest workmen and the frowns of men who're pure.

Onward, ye brave miners, keep steadily in sight
The future that's before us, the past so full of blight.
We'll strike against oppression, and guard our sacred cause,
And still be loyal subjects and keep our country's laws.

> Forward be our watchword, let us all unite,
> And bravely fight for Whitehouse, for Union and for Right.

The Durham Lock-out

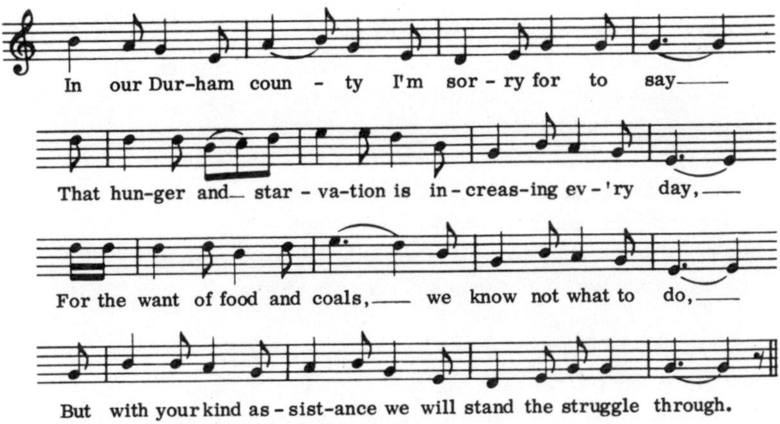

In our Durham County I am sorry for to say,
That hunger and starvation is increasing every day,
For the want of food and coals, we know not what to do,
But with your kind assistance we will stand the struggle through.

I need not state the reason why we have been brought so low,
The masters have behaved unkind, which everyone will know;
Because we won't lie down and let them treat us as they like,
To punish us, they've stopped the pits and caused the present strike.

May every Durham colliery owner that is in the fault,
Receive nine lashes with the rod, and then be rubbed with salt.
May his back be thick with boils so that he may never sit,
And never burst until the wheels go round at every pit.

The pulley wheels have ceased to move which went so swift around,
The horses and the ponies too all brought from underground,
Our work is taken from us now, they care not if we die,
For they can eat the best of food, and drink the best when dry.

The miner and his wife too, each morning have to roam,
To seek for bread to feed the hungry little ones at home.
The flour barrel is empty now, their true and faithful friend,
Which makes the thousands wish today the strike was at an end.

We have done our very best as honest working men.
To let the pits commence again, we've offered to them ten.
The offer they will not accept, they firmly do demand
Thirteen and a half per cent, or let the collieries stand.

Let them stand or let them lie or do with them as they choose,
To give them thirteen and a half we ever shall refuse.
They're always willing to receive, but not inclined to give.
Very soon they won't allow a working man to live.

With tyranny and capital they never seem content,
Unless they are endeavouring to take from us per cent.
If it was due, what they request, we willingly would grant;
We know it's not, therefore we cannot give them what they want.

The miners of Northumberland we shall for ever praise,
For being so kind in helping us those tyrannising days.
We thank the other counties too, that have been doing the same,
For every man who hears this song will know we're not to blame.

The Miners' Lock-out

Ye gallant lads of Lancashire, come, listen unto me.
I will unfold a tale of woe that's very sad to see.
Our children they are starving, you can see them day by day,
The offsprings of our collier lads for food they have to pray.

Then let us be united, we never must give way,
Uphold the Federation, lads, and we will win the day.

It's very hard on us poor lads that we must go away
To beg for our maintainance – we do it day by day.
But 'tis better far to do it than that we should engage
To go and take our shirts off and work for a pauper's wage.

But we must keep our tempers! Don't let our hearts go down,
We are getting well supported by the people of the town.
The publicans and the tradesmen throw in their little mite;
They are working well in our behalf, they know we're in the right.

We must thank our trusty leaders, they're well worthy of their steel;
The masters have not done what's right; the hunger they don't feel.
They have found their opportunity, it was not hard to seek.
We'd nothing [*broadside print illegible*] now we must put on the cheek.

All honour to Sam Woods, my lads, he's doing all he can,
Trying to get an honest wage for the British working man.
The day is fast approaching, when the victory we will shout
And we'll remember those who helped us when we were all locked out.

Don't forget the collier lads that are trying with their might,
Enduring so much suffering to get that which is right;
And when you see his box displayed no matter where he'll roam,
Think of his wife and children who are starving in their home.

Then let us be united, we never must give way;
Uphold the Federation, lads, and we will win the day.

Who Will Help Them?

Have you heard of the strife amongst our poor colliers?
It is caused by the masters who press them so hard;
Their families are dying of cold and of hunger,
And no one to help them to bear their hard lot.
    They asked their employers to give them more money,
    To provide for their families and let them have bread.
    They scorned these poor workmen and drove them away,
    Without any comfort or rise in their pay.

    So who then will help them to keep out on strike,
    That the employers may have time this lesson to learn,
    Altho' we're poor colliers, we're still British subjects,
    And born in old England, the land of the free.

The kind-hearted colliers of the north of Old England
Have offered to help them and so set them free
From their tyrant oppressors who have long been supreme,
And these Radstock colliers will never give in.
    There's no end of danger they're in down below,
    Doing their best to get us good coal,
    Yet still in this danger they're sorely oppressed
    By those who employ them for wages so low.

    So who then will help them to keep out on strike,
    That the employers may have time this lesson to learn,
    Altho' we're poor colliers, we're still British subjects,
    And born in old England, the land of the free.

The Dunkerton Carting Boys

Working so hard at the Dunkerton Pit,
The carting boys one day,
They met beside the way
And thought they'd have a say
About running the waggons for fifty yeards
For just one ha'penny.
We say: 'This game
Will never do for we.'

> For we're the Dunkerton carting boys,
> Just the Dunkerton carting boys,
> We're the Dunkerton carting boys,
> Oh, so true.
> Though we're poorly and plainly dressed,
> We're as honest as some of the best.
> We'll all stick together whatever the weather,
> The Dunkerton carting boys.

Now we ask for a little advance
Of just one ha'penny.
We mean to have the d,
Then we'll contented be.
But the manager said: 'You've too much now,
And you shan't have no more,'
We say: 'Oh, shan't we?
Then our haulin's o'er.'

> For we're the Dunkerton carting boys,
> Just the Dunkerton carting boys,
> We're the Dunkerton carting boys,
> Oh, so true.
> Though we're poorly and plainly dressed,
> We're as honest as some of the best.
> We'll all stick together whatever the weather,
> The Dunkerton carting boys.

The men and the boys of the Dunkerton Pit
Came out on strike, you see,
For just one ha'penny,
But they mean to have the d.
They won't go back till they've got the advance
And won the victory,
So, boys, join in
And sing this song with me:

> We're the Dunkerton carting boys,
> Just the Dunkerton carting boys,
> We're the Dunkerton carting boys,
> Oh, so true.
> Though we're poorly and plainly dressed,
> We're as honest as some of the best.
> We'll all stick together whatever the weather,
> The Dunkerton carting boys.

The Durham Miners' Big Meeting

Through Durham's famed city, a gay and joyous throng
With bands and banners flying gaily march along.
Of all other Unions, Durham's the most strong,
For it's shown at the Durham Big Meeting.
From every mining village and every colliery
Come the miners and their wives, once more old friends to see,
And this year on the race-course things lively they will be
At the Durham miners' Big Meeting.

> There's bands loudly playing and banners flying high
> As the Durham miners and their wives march along with joy.
> We want two shifts instead of three, will be the battle-cry
> Of the miners at the Durham famed Big Meeting.

From Blaydon, Dipton, Greenside, the Hobsons and the Spen,
Leadgate and High Stalks, South Moor and Annfield Plain,
Craghead, Burnhope and Kyo, and the Winnings here again,
And Crawcrooker's making straight for the Big Meeting.
From Ushaw Moor and Sleatburn, Cornsey and Langley Park,
The lads from Brandon and the Boyne will as usual have a lark,
And at night they'll sing *We're off again* to the Brownie in the dark
From the Durham miners' famed Big Meeting.

From Spennymoor and Auckland and Doggy, lots you'll see,
Framwellgate Moor and Sagerston, Kibblesworth and Pity Me,
Waldridge, Birtley, Chester-le-Street, also Edmondsley
With the Hettons will be at the famed Big Meeting.
From No Place and Grange Villa and Harden they are here,
The Sherburnes and the Pittings and Dean and Chapter near,
Clara Vale, Stargate and Whickham, and Winlaton's always near
To Swalwell at Durham's famed Big Meeting.

From Dunston and the Felling and Hebburn too,
From Hewarth, also Washington and Herrington they're through,
Houghton, Hetton and Hetton Downs, with Fence Houses ever true,
And the Lambtons, they'll be at the famed Big Meeting.
Whitburn, Ryhope and Monkwearmouth will be on Silksworth's track,
South Hetton too, and Shotton, good old Murton and the Nack,
From Trimdon, Wingate and Kello the trains they will be packed,
With Thornley close behind at the Big Meeting.

Backhall Mill, Chopwell, Willington, Crock and Byers Green,
From Shildon and the Blackhouse and Croxdale they have been,
From Morpeth and the Eldons and Sunnybrow they're seen,
With Newton cappers at the famed Big Meeting.
From Easington Lane and Consey, Haswell and Wheatley Hill,
Tanfield and Tantobie, White-le-Head and Rowlands Gill
Will hear of strong remarks made by a judge about their union's will,
And the three-shift system at the famed Big Meeting.

> There's bands proudly playing and banners flying high
> As the Durham miners and their wives march along with joy.
> We want two shifts instead of three, will be the battle-cry
> Of the miners at the Durham famed Big Meeting.

Miner's Life is like a Sailor's

Miner's life is like a sailor's,
'Board a ship to cross the waves.
Every day his life's in danger,
Still he ventures being brave.
Watch the rocks, they're falling daily,
Careless miners always fail;
Keep your hand upon your wages,
And your eyes upon the scale.
Union miners, stand together,
Do not heed the owners' tale.
Keep your hand upon your wages,
And your eyes upon the scale.

You've been docked and docked again, boys,
You've been loading two for one.
What have you to show for working,
Since your mining days begun?
Worn-out boots and worn-out miners,
And the children looking pale
Keep your hand upon your wages,
And your eyes upon the scale.

In conclusion bear in memory,
Keep this password in your mind:
God provides for every worker
When in union they combine.
Stand like men, and stand together,
Victory for you'll prevail,
Keep your hand upon your wages,
And your eyes upon the scale.
　　Union miners, stand together,
　　Do not heed the owners' tale.
　　Keep your hand upon your wages,
　　And your eyes upon the scale.

The Clydach Vale Lock-out

We are colliers from the Rhondda,
Locked out of work at Clydach Vale.
And it will surely make you wonder
When you've listened to our tale;
Since we had the new agreement,
Old customs they have tried to steal,
And though they had us at their mercy,
They have never made us squeal.

When they introduced Conveyors
Half our wages then they stole.
They say the pits aren't paying,
And they have thrown us on the dole;
Let the story of this outrage
Ring out over land and seas,
And may the kind South Wales coalowners
Know the want of bread and cheese.

Our kids are practically naked
Since the strike three years ago,
And with slack time and slump in wages,
We have never had a show;
What care they with all their millions,
In Guest, Keen and Nettlefold?
While we starve with unemployment
They are saving up their gold.

We have fathers, mothers, sisters,
Now in need this very day,
O kind friends, if you will help us,
God above will surely pay.

Farewell to Colliery Owners

For three hundred years or more you have owned the mines,
And all throughout those years, the miners had bad times.
You never cared, lived far away, the managers were your tools,
To make for you your profits, and acted more like fools.

Unlike the men, your nest is feathered. You do not need the dole.
But if justice reigned, you should now go down and hew the coal.
Your notices have expired now; of that we're very glad,
But of all men in the country, you'll be miserable and sad.

Farewell, you owners, we're sorry you haven't gone before.
You did much damage to our homes, and now you can't do more.
Farewell, farewell, the end has come, the sentence on you will be
'Inasmuch as you did it to the least ye did it unto me.'

I'm unfurling the flag today for the N.C.B. and state,
And hope everyone will do their best, as the scheme they undertake.
Be present every working day, that is if well and fit;
Then we shall hear the target fixed being reached at every pit.

The Pound a Week Rise

Come all of you colliers who work down the mine,
From England to Scotland, from Teesdale to Tyne,
I'll sing you a song of the pound a week rise,
And the men who were fooled by the government lies.
    And it's down you go, down below, Jack,
    Where you never see the skies,
    And you're working in a dungeon
    For a pound a week rise.

In nineteen and sixty, not three years ago,
The mineworkers' leaders to Lord Robens did go,
Saying: 'We work very hard; every day we risk our lives,
And we ask you here and now for a pound a week rise.'
    And it's down you go, down below, Jack,
    Where you never see the skies,
    And you're working in a dungeon
    For a pound a week rise.

Then up spoke Lord Robens. He made this decree:
'When the output rises, then I will agree
To raise up your wages, and give you fair pay,
For I was once a miner and I worked hard in my day.'
    And it's down you go, down below, Jack,
    Where you never see the skies,
    And you're working in a dungeon
    For a pound a week rise.

The miners went homeward. They worked hard and well,
Their lungs filled with coaldust in the bosom of hell.
The output rose by fifteen, eighteen per cent and more,
And when two years had passed and gone it rose above a score.
    And it's down you go, down below, Jack,
    Where you never see the skies,
    And you're working in a dungeon
    For a pound a week rise.

So the mineworkers' leaders went for their hard-won prize,
To ask Lord Robens for the pound a week rise.
But Robens wouldn't give a pound, he wouldn't give ten bob.
He gave them seven and six and said: 'Now get back to your job.'
    And it's down you go, down below, Jack,
    Where you never see the skies,
    And you're working in a dungeon
    For a pound a week rise.

So come all you colliers, take heed what I say,
Don't believe Lord Robens when he says he'll give fair pay.
He'll tell you to work hard and make the output rise,
But you'll get pie in the sky instead of a pound a week rise.

It's Miner this, it's Miner that

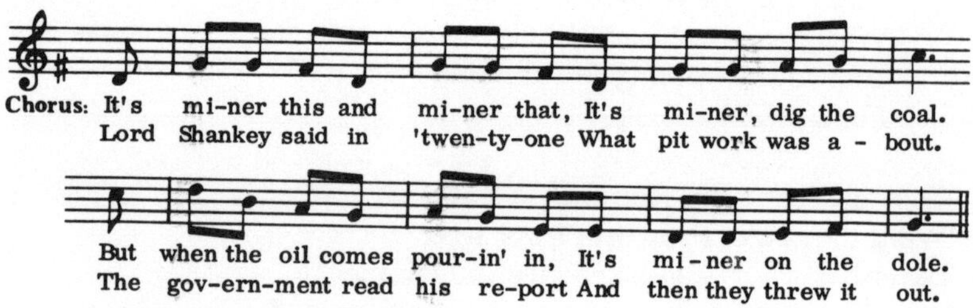

It's miner this, it's miner that,
It's miner, dig the coal.
But when the oil comes pourin in,
It's miner on the dole.

Lord Shanky said in 'twenty-one
What pit work was about.
The government read his report
And then they threw it out.

Remember what in 'twenty-six
Caused all the miners rage.
The masters wanted longer hours
And cut the pitmen's wage.

Alone they stood for eight long months,
No ally to be found.
'Twas then the law-and-order mob
Beat miners to the ground.

Then there was the good old days;
Depression, doom, despair.
Hardship was a way of life.
The miner got his share.

On battlefields of World War Two,
Dunkirk to Singapore,
There lies a million mining men.
They'll fight and die no more.

They say the miner has no dust,
He's only short of breath.
But, miner, in your sunken eyes
I see the living death.

People do not understand.
I wish they'd only try;
For I have seen the miner live,
I've seen the miner die.

> It's miner this, it's miner that,
> It's miner, dig the coal.
> But when the oil comes pourin in,
> It's miner on the dole.

Cheer up, Lads

Now they say the colliery's closin; don't you believe it, lads.
There's work left here for you an' me, as there was for our dads.
Now, the salvation of this colliery is the vaunted Busty seam,
Wi' a host o' bright officials an' some ace trepanner teams.

Now the seam height in the Busty is three foot six at most;
Why, on average that's a foot o' coal an' two foot six o' post.
The trepanner can cut this, an' he cuts it on the cheap –
All he needs, three sets o' picks a shift an' a new wheel twice a week.

Now, it's dampish in the trunk gate, as everybody knows,
But the lads have gills behind their lugs an' webs between their toes,
An' the howerman's a lovely payer, an' it's an even bet,
When ye go down for the third time, ye'll get paid excessive wet.

So tek yerself a three-piece suite on, or a new car, get the lot.
'Cause all of ye can rest assured this pit'll go till it stops.

Farewell to the Monty

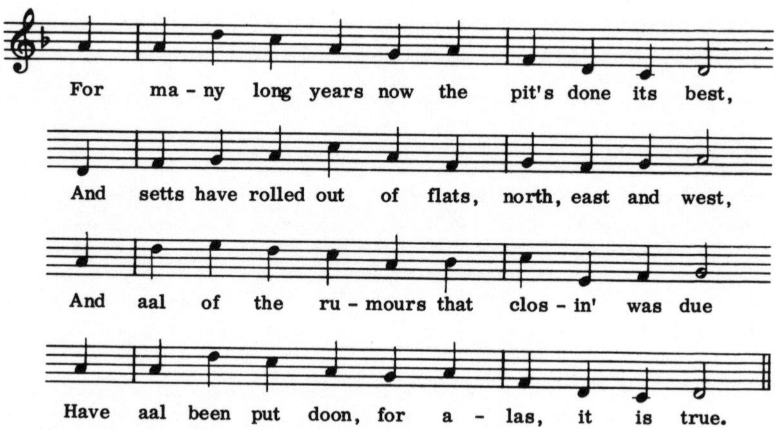

For many long years now the pit’s done its best,
And setts have rolled out of flats, north, east and west,
And aal of the rumours that closin’ was due
Have aal been put doon, for alas, it is true.

A meetin’ was held to discuss the affair,
And the manager said tiv us, right then and there:
‘We’ll have one last go before this pit is done,
To show a good profit on each single ton.’

But though profits were made, through the stocks pilin’ high,
The Coal Board decided the pit had to die.
An’ as output comes doon we get transferred away
To pits to the east for the rest of wor days.

Aa've filled in the Fan Pit, Aa've cut in the seam
In the Newbiggin Beaumont since Aa was fifteen.
Aa've worked at the Sections and in the Main Coal.
Man, it's hot doon the Monty, she's a dorty black hole.

So farewell to ye, Monty, Aa knaa yer roads well,
An' the work has been good an' the work has been hell.
Ne mair to yer dorty aad heap will Aa come,
For yer work is aal finished, an' yer life it is done.

Lament for Albert

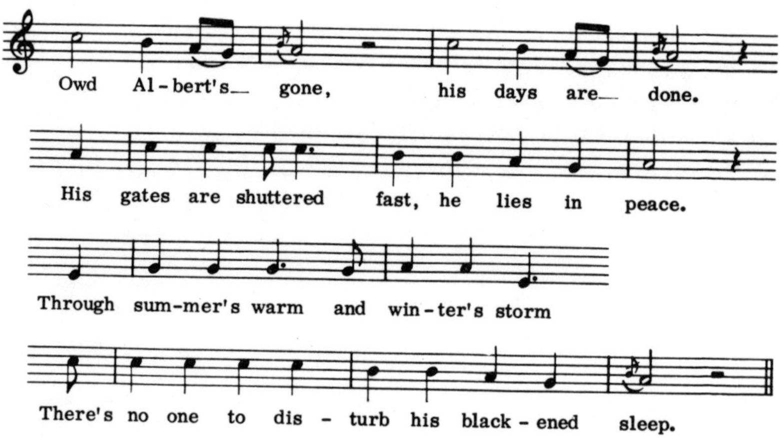

Owd Albert's gone, his days are done,
His gates are shuttered fast, he lies in peace.
Through summer's warm and winter's storm
There's no one to disturb his blackened sleep.

Four hundred men kept by hissen
Are gone away to work some other seam.
The tallies are hung, the knell is rung,
Lamps no longer through his darkness gleam.

No iron clogs or proppin'-logs
Move down the ramp a-shiftin' to the face.
And firedamp creeps and water seeps,
Long held at bay but movin' to their place.

Of twisted steel and broken wheels,
Owd Albert has them reckoned in his store.
Scattered around, they're all writ down,
Marked in the dusty ledger of the floor.

Closed is the door. He'll see no more
That hive of colliers workin' at the coal.
A million years, ten million fears
Sleep on inside the mine of Albert's soul.

Farewell to 'Cotia

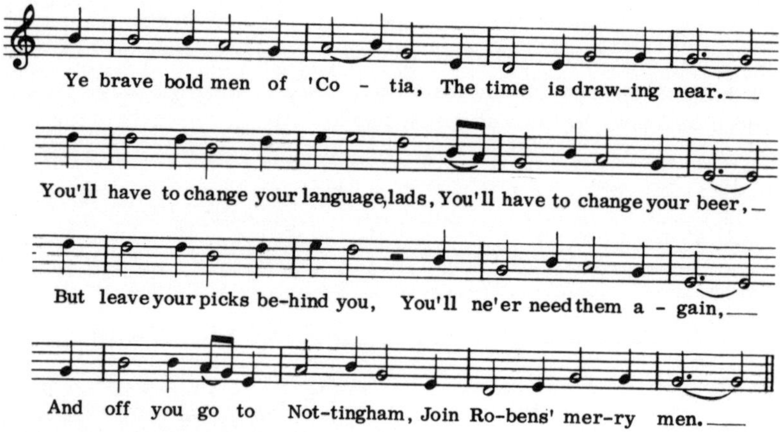

Ye brave bold men of 'Cotia,
The time is drawing near.
You'll have to change your language, lads,
You'll have to change your beer.
But leave your picks behind you,
You'll ne'er need them again,
And off you go to Nottingham,
Join Robens' merry men.

Ye brave bold men of 'Cotia,
The time is drawing thus.
You'll have to change your banner, lads,
And join the exodus.
But leave your cares behind you,
Your future has been planned,
And off you go to Nottingham,
To Robens' Promised Land.

Ye brave bold men of 'Cotia,
To you I say farewell.
And somebody will some day
The 'Cotia story tell.
But leave your cares behind you,
The death-knell has been tolled.
'Cotia was the colliery.
Her men were true and bold.

The 'Cotia Banner

Unfurl the 'Cotia banner, boys! Where are the rebels bold
That marched behind the banner on Gala days of old?
It was said that they were Red, blacklisted, I've been told.

> But there will be no more redundancy.
> Work for all in seams beneath the sea.
> For the AGM and all his men, hip, hip, hip, aree!

We'll mechanize old 'Cotia pit, machine-mad mech men said.
Even through the roof's alive, the coal we get is dead.
Have no fear, we'll shear and shear old 'Cotia out the red.

Drape the 'Cotia banner, boys, and hang your heads in shame.
Ten million tons are left behind, and who the hell's to blame
That men or mines don't matter in the economic game?

Fly the 'Cotia banner from the top of Bunker Hill.
Let them know old 'Cotia's dead, the pulley-wheels are still.
Then we'll march to Fatfield, say farewell to 'Cotia Bill.

Fare ye well, me 'Cotia lads, Topper, Bett and Tim.
Hope your cavil's canny where coal is still to win.
Remember, lads, to jowl the top and keep your timmer in.

> Oh, there will be no more redundancy.
> Work for all in seams beneath the sea.
> For the AGM and all his men, hip, hip, hip, aree!

Durham Big Meeting Day

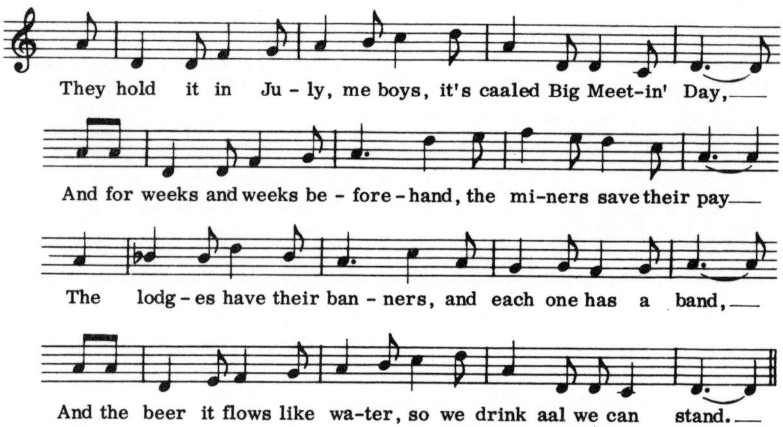

They hold it in July, me boys, it's caaled Big Meetin' Day,
And for weeks and weeks beforehand, the miners save their pay.
The lodges have their banners, and each one has a band,
And the beer it flows like water, so we drink aal we can stand.

The polises close the roads up, they torn the cars away
For Durham is stoppened off, me lads, upon Big Meetin' Day.
The bandsmen aal assemble with their horns so blindin' bright,
And they blaa from in the mornin' till gannin' hyem at night.

The crowds they aal gather an' the bands begin te play.
They march up to the Meetin' groond, an' Ryehope leads the way,
An' then we hear the speeches, an' they tell us what gans on,
Of wor hopes an' troubles, an' where the government's gannin' wrong.

Noo, as they speak of present times, me mind it wanders back
To work an' unions long ago when masters broke yer back.
Aa mind me aad grandfather, an' what he said te me
Of how they fought te get their dues, how unions came te be.

They broke the yearly bond an' the masters' crooked scales.
They fought the blackleg miners from Ireland and Wales.
When torned oot o' their hooses, wey, together they did stand
And they formed the Miners' Union that's known throughout the land.

So we've got better conditions and better money too,
We've taen the pits unto worselves to see what we can do.
But when we've paid the masters off with interest rates as well,
We'll show them coal's not gannin' back; we'll have a tale te tell.

Wey, with the speeches over, the folks aal torn aroond.
Te happins an' te bars an' pubs where fun can now be foond.
There's roondaboots an' shuggy-boats an' dodgem cars se fine.
An' lads an' lasses carryin' on, an' havin' a good time.

Noo, there wes some gannins-on, Aa think they should think shyem,
Aa blushed mesel when Aa went oot inte the back lane.
But then Aa thowt: 'Noo, shut thy gob, thoo's done the same thysel.
For if thoo thowt when thoo was young, thoo'd have a tale te tell.'

Noo, Aa's suppin' beer from dawn till dark, an' mind, Aa had me sup,
For Durham comes but once a year, an' it takes some savin' up.
Although there's fights an' arguments, Aa think we have a right
Te lowse worselves oot once a year, an' end up gettin' tight.

So if ye come wi' me next year, just get your pockets lined.
Fetch the missus an' the bairnies if ye have a mind.
We'll hear the band an' speeches too, an' spend maist o' wor pay,
But we'll have oorsels a bloody good time at the Durham Big Meetin' Day.

A Miner's Life

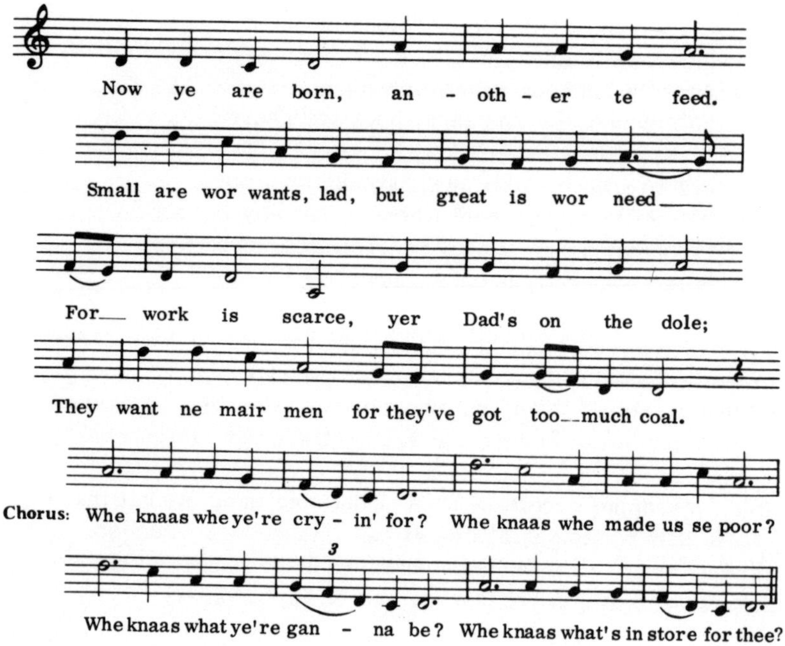

Now ye are born, another te feed.
Small are wor wants, lad, but great is wor need,
For work is scarce, yer Dad's on the dole;
They want ne mair men for they've got too much coal.

> Whe knaas whe ye're cryin' for?
> Whe knaas whe made us se poor?
> Whe knaas what ye're ganna be?
> Whe knaas what's in store for ye?

Now ye're a boy ye play in the streets,
Fish in the burn and performin' great feats,
While the pitheaps stand, markin' the graves
Of fathers and sons and the lives that they gave.

Now ye've left school, a man at thirteen.
Yer mother is grim as she looks on the scene
As she makes up yer bait; ye're off doon below,
And away te the pit wi' yer father ye go.

Now ye're a man, ye work in the mine.
Wi' fightin' and strugglin' ye're auld 'fore yer time,
For the work it is hard, conditions are bad,
And ye've promised yer son aal the things ye've not had.

Now ye are auld, ye've worked all ye can,
But they're closin' the pits and ye're part of a plan
To cut oot waste and make the mines pay.
So te hell wi' ye now, lad, and be on yer way.

VII

A Miners' Miscellany

VII

A MINERS' MISCELLANY

We have said that the miner is no different from any other working man when it comes to choosing songs. Most of what he sings has little to do with his work or his life, being the same popular pieces that are standard and current throughout the nation, with all their varying degrees of candour and bathos. Yet in some mining areas, particularly in the North-East of England, are many well-loved songs little known outside their own locality, that reflect the facts of life with humour, spirit and truth.

These homely ballads offer quite a different picture from the moth-eaten stereotypes of standard popular song with its panic or self-pity or nostalgia for the green hills of Athlone or Tennessee or some other region of the dream-world. Here instead is the rowdy sooty life of the row-town row with a pub at one end of the street and a pawnshop at the other, where pay day is slow in coming to the collier who has gambled his money away, and where mothers lull their babies to sleep under the shadow of hard times, to the refrain of 'pit lie idle'.

The row-town rows have not all disappeared, the pubs are still there, and the pawnshops may yet return. And though life may be more orderly and colliers less given to gambling and their wives never known to raise their voices over the backyard wall, the force of the old songs still remains. By chance, the most renowned of this kind of song is *Blaydon Races*, by George Ridley who went to work in the Oakwellgate Colliery at the age of eight. Wry comedies of pit village domestic life are presented in such pieces as *The Graveyard Shift* and *Wake up, wake up, my love*. But to hear a coalfields ballad at its sharpest and coaliest, one needs to be present when a hoarse-voiced Geordie is giving out with *The Row between the Cages*, the work, and probably the masterpiece, of Tommy Armstrong, the raffish pitman-poet of north-west Durham.

The Pitmen are not Bonny Lads

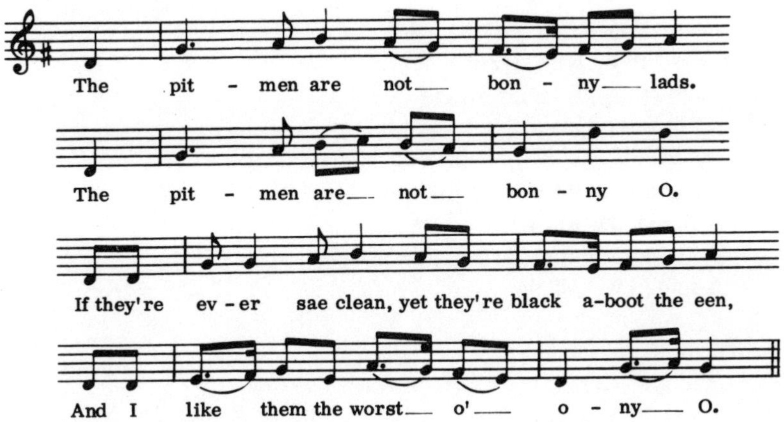

The pitmen are not bonny lads.
The pitmen are not bonny O.
If they're ever sae clean, yet they're black aboot the een,
And I like them the worst o' ony, O.

Tin Miners' Song

Oh, the farmers go around the fields,
Their legs tied up with straw.
The miners they go underground
And never miss a blaw.

> Oh, a-minin' we will go, my boys,
> A-minin' we will go,
> With picks and shovels in our 'ands
> A-minin' we will go.

Now I and Cap'n Franky
Got up to go to bal.
We started for Wheal Rodney
Where there was work for all.

Said I to Cap'n Franky:
What trebbut shall us 'ave?
Thirtain shellen and foorpence,
But foortain did us crave.

Us had fefteen foot of 'Saafety',
Of candles foortain pound,
And that was our materyal
For woorkin' underground.

Now us had luck at laast, boys,
The knackers shawed us wheer
To shut the rock and raise the tin,
And started us off feer.

Then us had jolly times, boys,
And plenty for to ait,
So us left a bit for Bucca
Who put us 'pon our fait.

We all worked at Wheal Easy,
And when the bal was scat,
We all lived in the engine house
And made the best of that.

Last Monday was Faste Monday.
We 'ad a bit to spend.
We took the maidens to the fair,
And now we've nought to lend.

We 'ad a sturt last Monday,
And ordered mutton stew,
Paid landlord Jan for nine months' beer,
And that was overdue.

And when we're up, we're up,
And when we're down, we're down,
And when we're only 'alfway up,
We're neither up nor down.

> And a-minin' we will go, my boys,
> A-minin' we will go,
> With pick and shovel in our 'ands
> We make a lively show.

The Drunken Exciseman

I know that young folks like to hear a new song
Of something that's funny and not very long.
It is of an Exciseman the truth I will tell,
Who thought that one night he was going to hell.
    Fal de lal.

One night he went out to look for his prey;
He did meet with some smugglers, as I heard them say,
In tasting the liquors they were going to sell,
The Exciseman got drunk, the truth I will tell.

He got so intoxicated he fell to the ground,
And like a fat sow was forced to lie down.
Just nigh to a coal pit the Exciseman did lie,
When four or five colliers by chance did come by.

They shouldered him up and hoist him away,
Like a pedlar's pack, without any delay;
Into the bucket they handed him down,
This jolly Exciseman they got under ground.

The Exciseman awakened with terrible fear,
Up started a collier, says: What brought you here?
Indeed Mr Devil, I don't very well know,
But I think I am come to the regions below.

Says the collier: What was you in the world above?
I was an Exciseman and few did me love;
Indeed Mr Devil, the truth I will tell,
Since I've got here, I'll be what you will.

Since you're an Exciseman, here you must remain,
You will never get out of this dark cell again;
The gates they are fast, and bind you secure.
All this you must suffer for robbing the poor.

Indeed Mr Devil, if you'll pity me,
No more will I rob the poor you shall see;
If you will look over, as you've done before,
I never will rob the poor any more.

Come give me your money, which now I demand
Before you can get to the Christian land.
Oh yes, Mr Devil, the Exciseman did say,
I wish to get back for to see light of day.

The Boomerang

Oh, listen a while and aw'll sing ye a sang
Aboot a strange thing that they ca' Boomerang;
It tummels aboot i' the air sae queer,
Like a pitman that's fuddled wiv ower much beer.
Chorus: Oh! Boomerangee,
Ah! Boomerangee,
It's 'stonished the folks,
Hes the awd Boomerang!

It's shaped like the busk in wor Malley's new stays;
It bends and it twists in a' sorts of ways;
The thrawer he sends it away wiv a bang,
But strange for to tell, back comes awd Boomerang.

Wor pit it laid idle the whole of the day,
To see the grand Fetee we bended wor way;
Thear was koiting and louping and strange things amang,
But ah canna forget that queer Boomerang.

Wor Malley she stared till her eyes were quite sair,
And 'Faither,' says Jacky, 'Its now gane for fair.'
'Hoot man,' says wor keeker, 'Thou canna be reet,
Its a new wooden sickle to maw down the wheat.'

The Indeans, they say, thraw'd to kill the white craws
But that seems to me agyane nature's awd laws;
For Newton, nor nayne ov the mathematical thrang,
In thear buiks say a word about awd Boomerang.

Wor sculemaister, he says its convex and concave,
It's curved and its pointed; the description he gave
Quite stonished wor brakesman as we cam alang,
The mathematical talk about awd Boomerang.

The beauties of Dryburn ah canna describe,
Wer ah poetickle it wad be me pride,
For the pitmen of Wear and Tyne divent knaw,
The fine sights and the sports that at Dryburn we saw.

Success to awd Wharton, and lang may he live,
Sic sports and sic pleasures the folks for to give;
And now that ah've very near finished me sang,
May Dryburn still flourish, and the awd Boomerang!
        Weel dune Wharton,
        Canny awd Wharton,
        Ye've 'stonished the folks
        Wiv the awd Boomerang!

Aw Wish Pay Friday wad Come

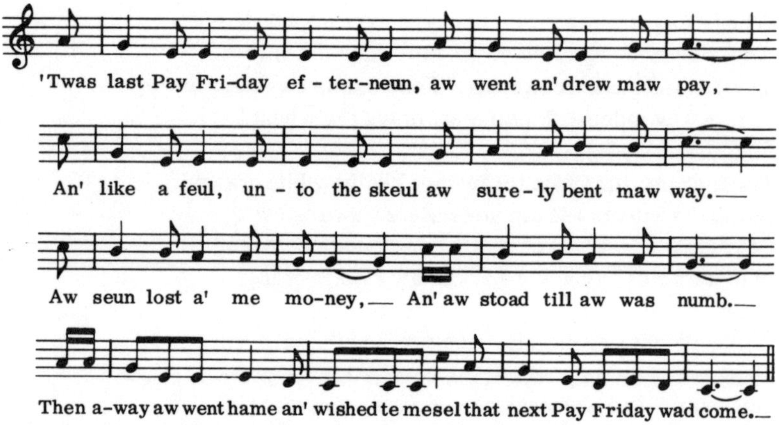

’Twas last Pay Friday efterneun, aw went an’ drew maw pay,
An’ like a fyeul, unto the skeul aw surely bent maw way.
Aw seun lost a’ me money, an’ aw stood till aw was numb,
Then away aw went hame an’ wished to mesel that next Pay Friday wad come.

When aw went hame an’ teld maw wife, she nearly broke her heart,
She says: ‘Maw lad, such wark as this is sure te myek us part.
Aw wadn’t cared if thou’d comed drunk wi’ strang beer, whisky or rum.
Aw wad tyen the rest an’ dyen the best till another Pay Friday wad come.’

Then she sobbed an’ sighed an’ the bairns a’ cried, an’ aw was varry bad;
A house confused an’ a woman’s abuse is enough te drive a man mad.
Aw knew varry weel what caused it all, so aw sat as if aw was dumb,
To speak aw was flaid, so naught aw said, but aw wished Pay Friday wad come.

The grocer an' butcher an' shoemaker tee, they a' come smilin in,
But what was maw poor wife to de but tell them she had ne tin?
Their smiles wes a' torned into frowns, it nearly struck them dumb,
An' when they went oot, aw couldn't say nought, but aw wished Pay Friday wad come.

On Saturday morn to be oot o' the way, aw took myself off to the town,
But hevvin ne brass to set me in, had to wander up an' down.
Aw met mony a kenned fyece in the street, but they all appeared to be dumb,
An' a' the way hyem, aw sang to mesel, aw wish Pay Friday wad come.

On Sunday morn when aw got up, the sun se bright did shine,
There was nought provided in the hoose to break wor fast or dine.
The bairns was cryin oot for broth, an' a greet marrow-byen made some;
They myed the house ring wi' tryin' to sing, aw wish Pay Friday wad come.

On Monday morn the miller cam in. Maw wife began to cry.
He said if he couldn't get his tin, he wad surely stop the supply.
Aw's proud to remark that aw was at wark, an' oot o' the way o' the hum,
An' a' the whole day aw was singin away, aw wish Pay Friday wad come.

We had nought to eat, neither taties nor meat, an' the bairns was cryin' for breed,
Maw wife was freetin away her life, an' aw wished that aw was deed;
Maw bran' new suit had gan up the spoot, it's a regular practice wi' some,
But not a good plan for a hard-workin' man, so aw wish Pay Friday wad come.

But next pay Friday aw'll lay maw life, aw'll not be such a fyeul.
Aw'll tyek my pay strite hyem to maw wife, i'stead o' gannin to skeul,
Aw'll treat mysel wi' a glass o' good yell, an' maw wife wi' a good glass o' rum.
An' aw'll give her the rest, to manage her best, so aw wish Pay Friday wad come.

Durham Gaol

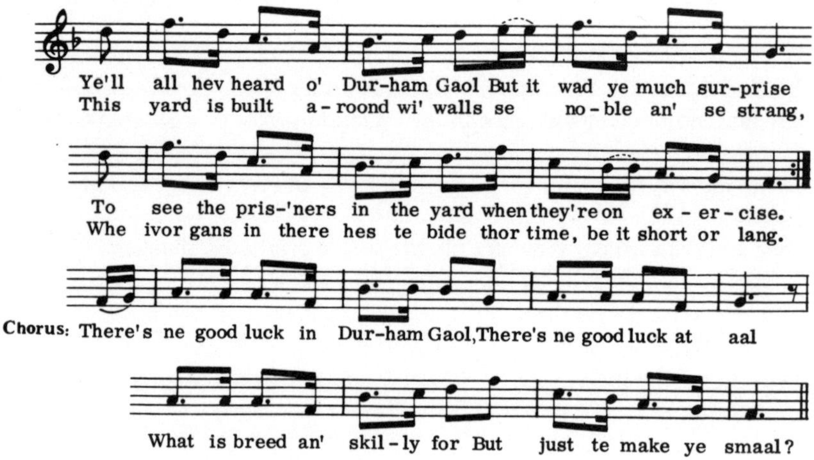

Ye'll all hev heard o' Durham Goal,
   But it wad ye much surprise,
To see the prisoners in the yard
   When they're on exercise.
This yard is built aroond wi' walls
   Se noble an' se strang,
Whe iver gans there hes te bide
   Thor time, be it short or lang.

      There's ne good luck in Durham Gaol,
         There's ne good luck at aal.
      What is breed an' skilly for
         But just te make ye smaal?

When ye gan to Durham Gaol
  They'll find ye wiv employ.
They'll dress ye up se dandy
  In a suit o' cordyroy;
They'll fetch a cap wivoot a peak,
  An' niver ax your size,
An' like your suit, it's cordyroy,
  An' comes doon ower your eyes.

The forst month is the warst iv all,
  Your feelins they will try,
There's nowt but two greet lumps o' wood
  On which ye hae te lie.
Then eftor that ye get a bed,
  But it is hard as stanes.
At neet ye dorsen't make a turn
  For fear ye break some banes.

All kinds o' wark there's gannen on,
  Upon them noble flats,
Teasin oakum, makin balls,
  An' weavin coco mats.
When ye gan in ye may be thin,
  But they can make ye thinner;
If your oakum isn't teased
  They's sure to stop your dinner.

The shoes ye get is often tens,
  The smallest size is nine;
They're big enough to make a skiff
  For Boyd upon the Tyne.
An' if ye should be caud at neets,
  Just make yoursels at hyem;
Lap your claes aroond your shoes
  An' get inside o' them.

Ye'll get your meat an' claes for nowt,
Your hoose an' firin free;
All your meat's browt te the door –
Hoo happy ye should be!
There's soap an' toowel, an' wooden speun,
An' a little bairnie's pot;
They fetch ye papers every week
For ye te clean your bot.

There's ne good luck in Durham Gaol,
There's ne good luck at aal.
What is breed an' skilly for
But just te make ye smaal?

The Row between the Cages

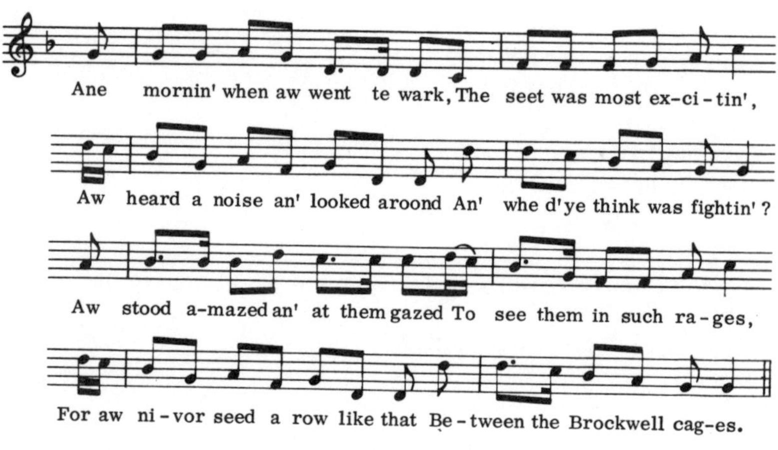

Ane mornin when aw went to wark,
The seet was most excitin,
Aw heard a noise an' looked aroond,
An' whe d'ye think was fightin?
Aw stood amazed, an' at them gazed,
To see them in such rages

For aw nivor seed a row like that
  Between the Brockwell cages.

Wor aud cage says: 'Come ower the gates,
  Because its my intention
To let thee see whether thoo or me
  Is the best invention.'
The new 'un bein raised, took off his claes,
  Then at it they went dabbin;
The blood was runnin doon the skeets
  An' past the weighman's cabin.

Wor aud cage says: 'Let's hae me claes,
  Thoo thought that thoo could flay me,
But if aw'd been as young as thoo,
  Aw's certain aw could pay thee.'
The patent knocked his ankle off,
  An' they both had cutten faces;
The shifters rapped through for to ride,
  So they both went to their places.

When gannin up and doon the shaft,
  The patent cage did threaten
For to take wor aud 'un's life
  If they stopped it meetin.
Wor aud cage bawled oot as it passed:
  'Thoo nasty dorty patent,
Rub thee eyes again the skeets –
  Aw think thoo's hardly wakenet.'

The patent to wor aud cage says:
  'Although aw be a stranger,
Aw kin wark me wark as well as thoo,
  An' free the men frae danger.
Noo, if the rope should break wi' me,
  Aud skinny jaws, just watch us –
Thoo'll see me clag on to the skeets,
  For aw's full o' springs an' catches.'

Wor aud cage to the patent says:
  'Aw warrant thoo think thoo's clever
Because they've polished thoo wi' paint,
  But thoo'll not last for ever.
The paint on thoo will wear away,
  And then thoo's lost thy beauty;
They never painted me at all,
  An' still aw've deun my duty.'

The brakesman brought them both to bank
  The mischief for to settle,
They fit frae five o'clock to six,
  An' the patent won the battle,
It took the brakeman half a shift
  To clag them up wi' plaisters.
Wor aud cage sent his notice in,
  Just to vex the maisters.

Scalloper Jim

Aa knew a chep called Scalloper Jim
(Ye seldom see a man like him),
He was queer shaped, but give him his due,
'Tis a weel knawn fact that he could hew.
And so he should, he was in his prime,
And he nivver lost a minute o' time.
To moist his hands, if them he'd lick,
Behint his lug he'd put his pick,
Just like a dole clerk dis wi' his pen
When writing, every now and then.

Big-muscled was he, but very small-brained.
He hardly had sense to come in when it rained.
I used to smile when he had need
At any time to scratch his heed,
For sure as life, my dearest friends,
He'd get a spelk in his finger-ends.

And weel Aa mind one windy neet
He lost his cap in Silver Street
And went in a shop to buy another,
And he didn't half cause the folks some bother.
They tried him in this, they tried him in that,
And then they tried a billycock hat;
But there wasn't a one in the place to fit,
And Aa think they might be trying yit,
But to oblige the poor feller,
They sewed him a peak on a big umbrella.

One day he'd had more than a drop,
And he fell close by a music shop.
His feet they went up near the roof
And darkened the windows down eneugh
That the shopman in surprise he mutters:
'Is someone putting up the shutters?'
So Scalloper, him to pacify,
Said a mouth-organ he would buy.
So in he goes, and they started laughing,
For his mouth was like a galloway's braffen.

So as they hadn't a one to suit,
He was just about thinking o' coming oot,
When the manager said: 'It would give much pleasure
If you'd have a special one made to measure.
It will cost you just about ten and a tanner,
So try your mouth on this grand pianner.'

When at the kist he was sitting at ease
He could beat a checkweighman telling lees.
I've heard him boasting many a time
Of them big ships that sail the Tyne;
He could fill one 'fore the putters come.
– Ye'd better hear that as be deaf and dumb!

He worked in a place called Camelphant Hill
Where he had to lie on his belly to fill,
And before he'd got his first tub full
He'd worn the shaft off a brand new shull,

For it was that laa that him and his mate
Each took a flat fish for his bait.

One morn in forst, his marrow was drilling,
And Scalloper he was slushing and filling,
When the whaff of his shull put out his leet,
And blew the deputy off his feet
And clashed him up against the pack,
For which poor Scalloper got the sack,
Which put an end to his hewing coal,
So now he's fattening on the dole.

spelk = splinter

Little Bit o' This

In a little black hole, they dig for the coal,
Me brother and his dear old dad.
They slave each day to earn their pay,
And nearly drive you blooming mad.

> Wi' their little bit o' this, and their little bit o' that,
> And their little plug o' twist to chew,
> And a little pair o' whirlers round their boots,
> What can the old woman do?

But on Saturday morn as sure as you're born,
She'll hold out her hand for his pay,
With a nice clean apron pinned up tight,
She'll get them ready for the next Monday.

> Wi' their little bit o' this, and their little bit o' that,
> And their little plug o' twist to chew,
> And a little pair o' whirlers round their boots,
> What can the old woman do?

Me Gaffer's Bait

Listen te me story an a tale Aa'll relate
Aboot the adventures of me gaffer's bait.
He'd fergotten his bag an he gied it te me,
There wes two home-made teacakes an a big flask o tea.

Now, doon in the pit ye'll meet a kinds o men,
There's canny folk, toonies an ones from the pen,
They might steal yer watch an yer lowse will be late,
But there's naethin more sacred than a miner's own bait.

Now, knaain hoo precious wes this cup o tea
Aa stowed it awey, as safe as could be,
Till when having a blaa doon at fowerteenth west,
Aa sat on me bait-bag, ye'll guess aal the rest.

Aye, Aa put me hand doon an his flask it wes broke,
An me gaffer torned red an Aa thought he wad choke,
An when he'd cooled doon an we went on wor wey,
Another thing happened for te ruin me day.

Now, the gaffer, that's Dixon, had stomped on aheid,
When Aa glanced at the way, an Aa saw somethin red,
So Aa stopped, rubbed me eyes, and torned aal aroond,
For there lay the gaffer's teacakes an jam on the groond.

There wes maa footprints on them, an those of my mate,
There wes no hope that day for my gaffer's bait.
As tremblin Aa told him, te mesel Aa did pray,
An he made us te suffer, he had my bait that day.

So take warnin from me if ye're given this task
Just watch what ye're deein, that's aal that Aa ask,
Ye can rob, ye can steal, ye can ride an 'oor late,
But when ye're warkin fer the gaffer, take care of his bait.

Hally's Piebald Gallowa

There wes a pit near Ches'-le-Street, they called it Lumley Six.
'Twas gettin' nigh Big Meetin Day, the lodge wes in a fix.
They hadn't lost the banner, but somethin' even worse.
I know you won't believe me, it wes eaten by a horse.

> The banner it wes grubby, noo, that wad never do.
> Lumley's wes the brightest the Gala ever knew.
> The lodge in aal its wisdom decided on the tub.
> Lumley wives wi' lovin care began to scrub an' rub.

They possed an' tossed the banner till it was bright again,
Once more fit for pride o' place o'er Lumley's marchin' men.
The possin' it wes over, they hung it oot to dry.
What happened to the banner made the noble Lumleys cry.

> They took it to the football field and hung it o'er a post.
> 'Wor banner is the brightest,' again wes Lumley's boast.
> But when they did collect it, 'twas tattered an' all torn.
> Some vandal there had been at work that fateful Monday morn.

The villain sure enough wes found, a four-legg'd fiend called Saul.
He wes a piebald gallowa belongin' Jimmy Hall.
The gallowa wes hungry, the banner tasted good.
It ate half o' Keir Hardie, and chewed up Martin Jude.

> The lodge it didn't panic, though there were sleepless nights.
> The lads still marched to Durham, there to claim their rights.
> But when they marched up Silver Street, folk couldn't understand
> Why Lumley Six was bannerless while marchin' with their band.

So let this be a warnin', if your banner's not so bright,
When you hang it up to dry, guard it day and night.
The poor old Lumley warriors, they never did forget
When Hally's piebald gallowa their bonny banner ate.

Ne Netties doon the Pit

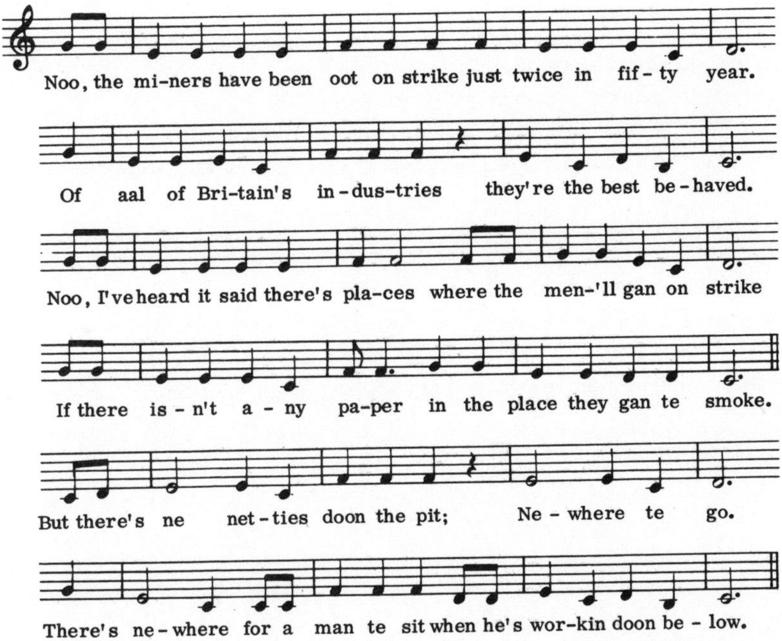

Noo, the miners have been oot on strike just twice in fifty year.
Of aal of Britain's industries they're the best behaved.
Noo, I've heard it said there's places where the men'll gan on strike
If there isn't any paper in the place they gan te smoke.

> But there's ne netties doon the pit, newhere te go.
> There's newhere for a man te sit when he's workin doon below.

They're diggin coal from undergroond an' far below the sea,
But where's a collier lad te gan whenever he wants a wee?
Dirty black an' stinkin holes wi' aal the latest gear,
But where's a collier lad te gan to get rid o' last neet's beer?

> For there's ne netties, etc.

A bloke came oot the other day fre Headquarters NCB;
We'd getten aboot halfway inbye, in great distress wes he.
The look that came upon his face wes a look of terrible hurt.
He says: 'Where d'ye go in a place like this when ye get caught short?'

Aa says: There's ne netties, etc.

When the filler farted, the putter he turned pale.
He felt the timber tremblin an' he got blawn off the way.
'Had on,' says his marra, 'Ye'll bring the whole place doon!'
He says: 'Aa canna help it, Bob, aa've been oot on the Broon.'

An' there's ne netties, etc.

At lowse ye ride an' come te bank at the end o' every shift;
A smoke, a bath an' a change o' clothes, an' then a little sit.
For them of you whe work at bank, do ye ever think it's great
That ye can gan an' wash yer hands afore ye get your bait?

'Cause there's ne netties, etc.

The Graveyard Shift

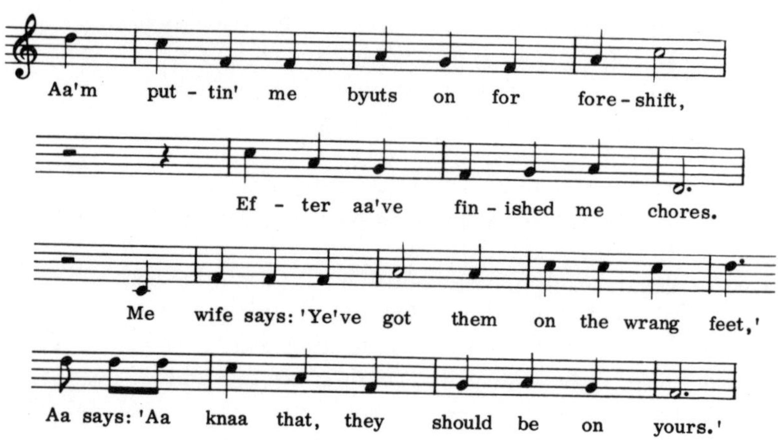

Aa'm puttin me byuts on for fore-shift,
Efter aa've finished me chores.
Me wife says: 'Ye've got them on the wrang feet.'
Aa says: 'I knaw that, they should be on yours.'

Aa pick up me bait an me bottle,
Ootside it's rainin se hard.
She says: 'It's not fit fer a dog to be oot,'
As she kicks us into the back yard.

> So let's drink te the graveyard shift, me lads,
> It's easy when the snaw's blaain,
> To say te yer marras: 'Te hell wi' the pit,
> Aa'll stop in the club, aa'll not gan.'

Noo, when aa get back from the forst 'un,
Aa'm that tired, aa cannot lift a cup.
Aa drift off to sleep, but she breaks me dream deep
An' says: 'What time will aa wake ye up?'

Noo, the wark's that much harder in fore-shift,
It's fower days a week he's te dee,
An' the only reason that aa wark fower
Is wor lass cannot manage on three.

> So let's drink te the graveyard shift, me lads,
> It's easy when the snaw's blaain,
> To say te yer marras: 'Te hell wi' the pit,
> Aa'll stop in the club, aa'll not gan.'

Noo, ma wife, aa caal her the Dragon.
Ye might think that's goin' a bit far.
Aa've got a good reason for caalin her that –
'Case she keeps draggin me oot o' the bar.

Noo, when aa come hyem carryin her Guinness,
The lads say that aa'll have te stop.
When aa tells 'em it's a bottle aa got fer the wife,
They say: 'What a bloody good swap!'

> So let's drink te the graveyard shift, me lads,
> It's easy when the snaw's blaain,
> To say te yer marras: 'Te hell wi' the pit,
> Aa'll stop in the club, aa'll not gan.'

After a Pit Pay Day

After the roar of a pit pay night,
Just at the break of day,
In all directions on the field
The noble heroes lay.

Some were too full to venture home,
Some could not find the way,
And all this world seemed upside down,
After a pit pay day.

Wake up, wake up, my love

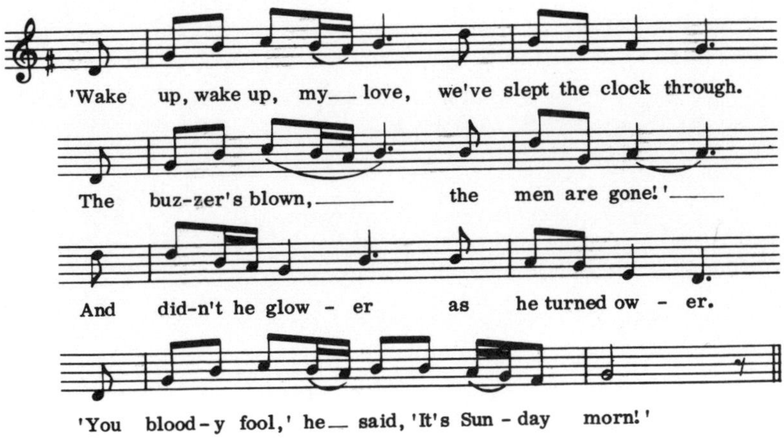

'Wake up, wake up, my love, we've slept the clock through.
The buzzer's blown, the men are gone!'
And didn't he glower as he turned ower.
'You bloody fool,' he said, 'It's Sunday morn!'

Pit Lie Idle

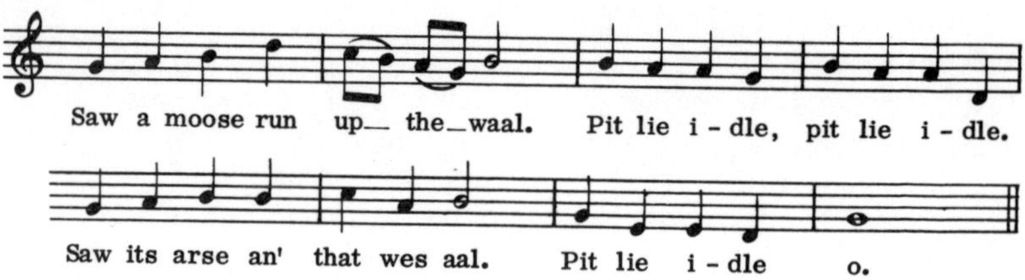

Saw a moose run up the waal.
Pit lie idle, pit lie idle.
Saw its arse an' that wes aal.
Pit lie idle o.

Notes

NOTES

THE COLLIERS' RANT. Our version, from the singing of John Elliott, Birtley, Co. Durham (31 July 1963), differs only in small detail – apart from the closing couplet – from the 'standard' set first published in Joseph Ritson's *Northumberland Garland; or, Newcastle Nightingale* (1793), and subsequently reprinted by various editors of North-Eastern song collections. The melody was first published in Robert Topliff's *Selection of the most popular Melodies of the Tyne and the Wear* (London, ?1815).

CALL THE HORSE, MARROW. From a made-up book of broadsides and other printed scraps, put together by John Bell (1783–1864), and now in the University Library, Newcastle-upon-Tyne. No imprint or date, but apparently issued late in the eighteenth century. The fourth stanza seems to envisage a similar accident to that which occurred later at New Hartley in 1862, when an engine beam broke and fell, blocking the shaft and causing many deaths.

JACKY IN THE LAW RAW. From a manuscript collection of local songs, partly assembled by John Bell *c.* 1830, in the University Library, Newcastle-upon-Tyne.

THE MINER'S LIFE. Text from E. D. Williams, collier, Brymbo, Denbigh, North Wales (1951).

THE COLLIER. From a broadside published by Wrighton of Birmingham. Quoted here from Roy Palmer (ed.), *Poverty Knock* (London, 1974). Palmer suggests that the song probably dates from the late 1820s. 'George our great king' was presumably George IV, who died in 1830. The firm of Wrighton flourished *c.* 1810–30.

THE COLLIER LASS. From a broadside published by John Harkness of Preston, *c.* 1850–55. One is reminded of the evidence given by seventeen-year-old Patience Kershaw to the Children's Employment Commission, Mines (1842), quoted in E. P. Thompson's *The Making of the English Working Class*: 'The bald place upon my head is made by thrusting the corves . . . I hurry the corves a mile and more under ground and back; they weigh 3 cwt. . . . the getters that I work for are naked except their caps . . . sometimes they beat me, if I am not quick enough.'

THE BOLD MINER. Text from Joseph Spooner, Fence Houses, Co. Durham. Remembered from the singing of his father-in-law, 'as he rocked the bairns on his knee in the old rocking chair'. Other versions, more or less identical, begin: 'Only a miner who works underground.'

DOWN IN A COAL-MINE (I). Text from a broadside published by H. Disley, 57

High Street, St Giles's, London, *c.* 1865. The song was re-made for stage performance by J. B. Geoghegan in 1872. See George Korson, *Minstrels of the Mine Patch* (Philadelphia, 1938), pp. 277–8, for an American version. Korson calls it 'the best-known miner's ballad'.

DOWN IN A COAL MINE (II). Text mainly from George Bailey, ex-miner, of Bentley, Doncaster (May 1951). Tune and missing fragments from James Hedley, of Aberavon.

FIVE IN THE MORNING. From a broadside published by H. Such, 177 Union Street, Borough (London) SE, in the Sheffield University Library's Collection.

COLLIER'S SONG. From a broadside (n.d.) printed by Bebbington, 26 Goulden Street, Oldham Road, Manchester, sold by H. Andrews, 27 St Peter Street, and John Beaumont, 176 York Street, Leeds.

THE COLLIER LAD IS A CANNY LAD. Written by 'Johnny Handle' (John Pandrich), of Winlaton Mill, Co. Durham, and transcribed from his singing on a Topic record, *The Colliers' Rant* (TOP 74).

THE FORE SHIFT. Written by Matthew Tate, 'of an old mining family' of Benton and Blyth, Northumberland. Printed in the 'Newcastle Weekly Chronicle' in 1886, and reprinted in Allan's *Tyneside Songs* (Newcastle-upon-Tyne, 1891).

THE HASWELL CAGES. From a broadside published by George Walker, Sadler Street, Durham, *c.* 1840. The prescribed tune is *The Wedding of Ballyporeen.* The colliery village of Haswell lies some nine miles south of Sunderland.

THE WINDER. Text from Michael Ward, miner, of Kirkby-in-Ashfield , Notts.

THE CAGE. Written by Keith Roberts, of Wigan, and published by him as a broadside, *c.* 1967. Tune transcribed from Mr Roberts's singing on a Trailer record, *Pier of the Realm* (LER 3031).

THE DEPTH O' THE MINE. Sung by Charles Bevil, shot-firer, of Fir Tree Drift, Tow Law, Co. Durham (June 1951).

THE DRIVER. From the singing of Rees Elliott, Birtley, Co. Durham. The words are by Alexander Barrass, from his *Pitman's Social Neet* (Consett, 1897). Mr Elliott sings the song to the tune of *The Mountains of Mourne.* He keeps fairly close to Barrass's text but we present this transcription because his touch with the dialect is lighter than Barrass's. 'Siss' and 'Baa (or Baw)' were pony-drivers' calls meaning 'Go' and 'Stop'.

I AM A DRIVER. Collected by A. E. Green from a Castleford miner, William Hill, in 1966.

THE STONEMAN. Written by 'Johnny Handle' (John Pandrich) and transcribed from his singing on a Topic record, *Stottin' doon the Waall* (TOP 78).

THE PICK-SHARPENER. Written by Michael Ward, of Kirkby-in-Ashfield, Notts.

OAKEY'S KEEKER. Written by Thomas Armstrong (1848–1919), the mine-balladeer of Tanfield Lea, Co. Durham. Published in W. H. Armstrong (ed.), *Song Book containing 25 Popular Songs of the late Thomas Armstrong* (Chester-le-Street, 1930).

THE PUTTER. A song by Alexander Barrass, published in his *Pitman's Social Neet* (1897). To the tune of *Wait for the Waggon*. Here transcribed from the singing of 'Johnny Handle' (John Pandrich) on a Topic record, *The Bonny Pit Laddie* (TOP 12TS 271).

THE PONY PUTTER. Written by George ('Jock') Purdon, of South Pelaw, Co. Durham, to the tune of *The Lambton Worm*.

THE VOLUNTEER PUTTER. From the singing of Rees Elliott, Birtley, Co. Durham (January 1973). A parody of the Victorian parlour ballad, *The Volunteer Organist*, by W. B. Gray. The song derives its force from the fact that 'putting' – conveying empty tubs to the face and taking away the full ones, either by means of a pony or by manhandling – is a particularly strenuous and disagreeable job, usually done by young men.

THE DOGGIE PUTTER. From the singing of Bert Draycott, of Fishburn, Stockton-on-Tees (June 1973), who says: 'I got this from an old pump man at Fishburn Colliery. He had been a deputy at West Cornforth Colliery before it was closed. His name was Walter Leather. West Cornforth is always called "Doggie" by the local people. I've never found out why.'

AA WES GANNIN INBYE. From the singing of John Elliott, Birtley, Co. Durham (July 1963). A parody of the 1930s hit, *Moonlight Bay*.

THE PUTTER'S LAMENT. From the singing of Bert Draycott, of Fishburn, Stockton-on-Tees (August 1974). Mr Draycott reported: 'I heard this one from a Welsh bloke about two years ago. You're supposed to hear this when you're walking inbye in back-shift and the fore-shift men are properly chewed off.' The sorrows of the putter whose full tub has come off the rails provide an endless source of lyrical comment.

POOR PUTTER. Likewise from the singing of Bert Draycott, as above. 'This one I got from a bloke at Bishop Middleham. A little puttin' ditty about nothing much. Poor old putter; they're all the same – inbye, outbye, off the way, all day.'

LITTLE CHANCE (I). As sung by Charles Bevil, shot-firer, of Fir Tree Drift, Tow Law, Co. Durham (June 1951). 'This is the song we used to sing as lads at Christmas time in our pit village, whilst going from door to door with the "Tup", fifty years ago.' In a letter (22 November 1963), Jack Elliott, of Birtley, Co. Durham, wrote of the custom of luck-visiting with the 'Tup' or 'Ram': 'I got this information recently from a friend of mine at the Workmen's Club. His grandfather told him that when he was a child at Cornsay, West Durham, weeks before Christmas they used to stand at the pit-head or drift-mouth and cadge all the pieces of candle the pitmen had left. These were saved, and on Christmas Eve they were placed all round a bier or cradle. A stiff piece of cloth was bent over to form a sort of covered waggon effect. With the handles of the bier sticking out like four legs it looked like an animal. That was the Tup. There was a trough at each end of the bier, which was used to collect coppers in when the Tup was paraded round the village. I remember these cradles, or creddles.

There was about four in our street. We had one. They were used for pig-killing operations.'

In former times, the Tup had special significance in the pit itself. On the last working day of the year, the last corf of coal was drawn ceremoniously out of the pit, with lighted candles all round and a ram's horns, real or represented, set on top of the basket. There is a suggestion that the ceremony symbolized the withdrawal of the generative force of the pit which had, at least temporarily, ceased production.

A variant of Mr Bevil's song was collected by W. Toyn from J. G. Piggins, of Consett, Co. Durham, who remembers it being sung in the Derwent Valley collieries, *c.* 1924. His song begins: 'Oh, my nyem is Archie Ramsden. Aa myed me fyece advance. Aa'm drivin' a little gallowa'. His nyem is Little Chance'. A fragmentary version was noted by Maxine Baker from Mrs A. Leyland who learnt the song as a child in Tow Law and, like other informants, recalled that there were several verses about various workers in the pit. Maxine Baker also noted a five-verse set from an ex-miner who learnt the song in Lambton 'D' pit, *c.* 1915.

LITTLE CHANCE (II). From the singing of John Elliott, Birtley, Co. Durham (July 1963). This version, which has become broadly popular since the middle '60s, has been much affected by a pit-pub song familiarized by the ex-miner comedian Billy Hutchinson, of Bedlington.

JOWL AND LISTEN. Collected by W. Toyn from Henry Nattress, of Low Fell, Co. Durham, in 1962.

EARLY IN THE MORNING. Text from H. Sherratt, Wheatley, Doncaster (May 1951).

WALKER PITS. First printed in John Bell's *Rhymes of Northern Bards* (Newcastle-upon-Tyne, 1812, facsimile reprint, 1971). Directed to be sung to the tune of *Off She Goes* (see, for instance, *Kerr's First Collection of Merry Melodies for the Violin*, Glasgow, n.d. Book 1, p. 29 No. 14). Other melodies have been attached to these words of recent years, including the tune of the American camp-meeting hymn *Where are the Hebrew Children?* (see *The Southern Harmony and Musical Companion*, New Haven, Conn., 1835, p. 266), and a version of the north-eastern dance tune *My dearie (laddie, lassie) sits ower late up.*

THE COLLIERS' PAY WEEK. Written by the printer Henry Robson, of Benwell (1775–1850). Text slightly modernized from Bell's *Rhymes of Northern Bards.*

WRECKENTON HIRING. Anonymous text from *A Collection of Songs Comic, Satirical and Descriptive, Chiefly in the Newcastle Dialect*, published by John Marshall, Newcastle-upon-Tyne, 1827. An undated broadside issued by W. Stephenson of Gateshead prescribes the tune: *The Flowers of Caledonia.*

BAGGY NANNY, OR THE PITMAN'S FROLIC. Words by the printer and dialect song-writer Robert Emery (1794–1871) from Fordyce's *Newcastle Song Book* (1842). The song arose from the visit of the violinist Niccolò Paganini to

Newcastle in September 1833. It is directed to be sung to the tune of *The Kebbuckstane Wedding.*

THE KIPPERED HERRIN'. Words by the former sail-maker and stage comedian Edward Corvan (1830–1865), from Allan's *Tyneside Songs* (Newcastle-upon-Tyne, 1891).

THE COCK-FIGHT. From the singing of James Hamilton, of Hunslet, Yorks, (March 1951). Other versions of this song are in Harland's *Ballads and Songs of Lancashire* (London, (1865), Ashton's *Modern Street Ballads* (London, 1888), Kidson's *Traditional Tunes* (Oxford, 1891), and Vaughan Williams and Lloyd's *Penguin Book of English Folk Songs* (Harmondsworth, 1959), from *Journal of the Folk-Song Society*, No. 6 (1905). The various versions locate the cock-pit in sundry places in Yorkshire, Lancashire and the Midlands.

SHOVELLIN' BACK THE SLATE. From the singing of Jack Elliott, of Birtley, Co. Durham, July 1963. A north-eastern re-make of a song originally composed by an Irish miner, Ed Foley, in Pennsylvania in 1892. It was brought to Durham about the time of the First World War by a pitman from Kentucky, Yankee Jim Roberts, with whom Jack Elliott's elder brother Rees 'worked marras'. During the song's stay in Durham, both text and tune have become much altered from the original as printed in George Korson's *Minstrels of the Mine Patch* (Philadelphia, 1938). The song is also known as *The* (or *A*) *Celebrated Working Man*, and *In the Bar-Room.*

STOTTIN' DOON THE WAALL. Written by 'Johnny Handle' (John Pandrich) and transcribed from his singing on a Topic record, *Stottin' doon the Waall* (TOP 78).

THE PITMAN'S UNION. From a broadside in the University Library, Glasgow. No imprint or date, but probably 1840s.

BRAVE COLLIER LADS. From a broadside, n.d. published by W. and T. Fordyce, Newcastle-upon-Tyne.

A PITMAN'S LOVE SONG. From a collection of local songs mainly assembled by John Bell about 1830, now deposited in the University Library, Newcastle-upon-Tyne. A short version of the song appears in Cuthbert Sharp's *The Bishoprick Garland* (London, 1834). The title is apparently Sharp's.

THE BONNY PIT LADDIE (I). Text and melody first printed in J. Peacock's *Favourite Collection of Tunes with Variations Adapted for the Northumbrian Small Pipes, Violin or Flute*, Newcastle, n.d. (*c.* 1800).

THE BONNIE PIT LADDIE (II). From the singing of Pat Elliott, recorded at Birtley, Co. Durham (31 July 1963). Mrs Elliott learnt the song from a collier's elderly widow, Mrs Hannah Short of Fatfield, Co. Durham.

THE COLLIER LAD. Text from H. Sherratt of Wheatley, Doncaster (July 1951).

COLLIER LADS. Text from Dave Hurst, Hindley Green, Wigan. Learnt by his father from a Hindley Green woman, when he was a boy. Mr Hurst said: 'My father sings the first two verses to a slight variant of *Johnny Todd*, and the final one to a tune I never heard before.'

I'LL HAVE A COLLIER FOR MY SWEETHEART. Text from William Oliver, of Widnes. Another, more fragmentary version was communicated by an ex-miner of Cwmynyscoy, Pontypool, formerly of Platt Bridge, near Wigan, who says the song was 'popular some fifty years ago among the collier boys and the girls who worked on the screens, and why the bit came in about the factory lads was because there were plenty of cotton mills around there at the time.' The song has some relation to the well-known *Johnny Todd* through common possession of the ring and cradle verses.

THE WAGGONER. Often reprinted in various forms since its first appearance in Bell's *Rhymes of Northern Bards* (1812). About the time that he was putting together the *Rhymes*, Bell was also compiling a tune-book, which remains in manuscript. It contains the earliest copy of the melody of *The Waggoner* (there called *Down the Waggon Way*) with a pencilled note identifying the tune as *The wind that blew the barley down.*

THE SANDGATE GIRL'S LAMENTATION. The words first appeared in Bell's *Rhymes of Northern Bards.* The first publication of the tune was in Joseph Crawhall's *A Beuk o' Newcassel Sangs* (Newcastle-on-Tyne, 1888) where it is identified as a variant of the air most commonly associated with *The Manchester 'Angel'.*

THE COLLIER LADDIE. Text communicated by Hamish Henderson of Edinburgh (1951). Robert Burns sent a fragmentary copy (five verses) to James Johnson, editor of the *Scots Musical Museum*, with the note: 'I do not know a blither old song than this.' A version with tune, noted about the beginning of the present century, appears in John Ord's *The Bothy Songs and Ballads of Aberdeen, Banff & Moray, Angus and the Mearns* (Paisley, 1930).

JIMMY'S ENLISTED, or THE RECRUITED COLLIER. Text from J. T. Huxtable, of Workington. A more literary version appears in R. Anderson's *Ballads in the Cumberland Dialect* (Wigton, 1808). Over the last quarter-century the song has had fair currency, sung to a tune fitted by myself.

JOLLY JOE THE COLLIER'S SON. From a broadside issued by H. Wadsworth, of Birmingham. Swindells, the Manchester printer, issued a similar version very early in the nineteenth century, as did J. Pitts of Seven Dials, London, under the title *Jack of Ambrose Mill.* Another version was issued by Stephenson of Gateshead, a little later. H. P. Such re-issued Pitts's copy *c.* 1870. Cecil Sharp, H. E. D. Hammond and George Gardiner found the song current in the South and West of England in the present century, with the place-names altered (sometimes but slightly – 'Holberry' instead of Oldbury, as on the Pitts and Such broadside, 'Bilson' instead of Bilston, etc.). For all its wide distribution, the ballad seems properly to belong to the Midlands. The tune we give was collected by Cecil Sharp in 1908 from John Bingle, of Lew Trenchard, Devon. It is No. 281 in Sharp's MSS. Mr Bingle called the song *Joe, the Penn Collier's Son.* Penn is a suburb of Wolverhampton, once an outlying village.

THE SANDGATE LASS ON THE ROPERY BANKS. Written by the blind fiddler,

Robert Nunn (1808–53). Words reproduced here from Allan's *Tyneside Songs* (1891), where the song is directed to be sung to the tune of *The Skipper's Wedding*. A setting is printed in J. Stokoe and S. Reay, *Songs and Ballads of Northern England* (Newcastle-on-Tyne and London, n.d./probably 1892). There, the tune-title is given as *Fy, let us a' to the Bridal*.

SANDGATE WIFE'S NURSE SONG. Another of Robert Nunn's songs, printed in Fordyce's *Newcastle Song-Book* (1842). The tune is that of the camp-follower song, *Dol-li-a*, first printed in Robert Topliff's *Selection of the most popular Melodies of the Tyne and Wear* (London, n.d./probably 1815). The present version, transcribed from the singing of Audrey Thorogood at Birtley, Co. Durham (16 March 1971), with its slight departures from the printed text, exemplifies the form in which the song is currently sung.

THE RAMBLING MINER. From a broadside without imprint or date. Another text, almost identical with this, likewise without imprint, is among the broadside collection in Sheffield University Library. The song seems to have started out in the mid-eighteenth century as *The Rambling Soldier*. In the 1820s the broadside printer James Catnach published an influential version, re-cast as *The Rambling Sailor*, which was reprinted several times by other broadside houses and was still widely current in the southern counties in the early years of the twentieth century (Cecil Sharp alone collected eight versions). *The Rambling Miner* is a transparent remake of *The Rambling Sailor*, lightly localized in the tin-mining country of Cornwall ('Mertha' is doubtless Merther, a few miles east of Truro). All the recorded versions of *The Rambling Sailor* were sung to variants of the same melody – a hornpipe tune, says Baring-Gould – and probably *The Rambling Miner* was also sung to some such melody as appears on pp. 305–310 of *Cecil Sharp's Collection of English Folk Songs*, ed. Maud Karpeles, London, 1974.

CUSHIE BUTTERFIELD. Words by the ex-collier comedian George Ridley (1834–64). Directed to be sung to the tune of *Pretty Polly Perkins*.

SIX JOLLY MINERS (I). As sung by Mrs Cosgrove of Newtongrange, Midlothian (11 May 1951), with additional stanzas from the version of Charles McVey, of Sanquhar, Dumfriesshire. Mr McVey remembers 'some of the men that worked in the same mine as my father coming to the house on a Saturday night, and this was their main song. I think the song has a very deep Lanarkshire origin.' For all that, the song has been widely diffused elsewhere, its spread stimulated by its appearance on a number of broadsides. Versions have been recovered from localities as far apart as Yorkshire, Hampshire, Nova Scotia and Pennsylvania, this last from an English migrant who learnt the ballad in her native Cumberland while still a little girl. Her version mentions (Bishop) Auckland, Newcastle and Durham (see G. Korson, *Minstrels of the Mine Patch*, Philadelphia, 1938, p. 271). Another of Korson's versions comes from a singer in Glace Bay, Nova Scotia. Mrs Cosgrove, who had the song from childhood, spent some years in Glace Bay, where she often sang it at miners' concerts. She remarks: 'Besides working in the mines in Nova Scotia, my husband has also been in the mines in Whitburn and Lanarkshire, so you see I think we followed our song.'

SIX JOLLY MINERS (II). From the singing of Francis Cudmore, of Ballycastle, Antrim; recorded in London, May 1974.

MY COLLIER LOVER. Written by Ernest D. Williams, miner, of Brymbo, near Wrexham, and communicated by him, June 1951. An unusually conventional piece for a pitman's composition. This kind of song more often emerges as a burlesque.

MA LAD'S A CANNY LAD. From the singing of John Elliott, Birtley, Co. Durham (July 21, 1963). To the tune of *The Ball o' Kirriemuir.*

PICK, PICK, PICK. As sung by S. A. Wharmby, of Skegby, Notts (1951), who said the song was 'very popular sixty years ago at weddings and parties'.

FOURPENCE A DAY. Collected by Joan Littlewood and Ewan MacColl from John Gowland, a retired lead-miner of Middleton in Teesdale. The song is attributed to Thomas Raine, lead-miner and bard of Teesdale. Communicated by Ewan MacColl.

LITTLE PAL. From the singing of Rees Elliott, Birtley, Co. Durham. To the tune of the sentimental popular song *Little Pal.* In another version, from Joseph Spooner, Fence Houses, Co. Durham (5 May, 1951), the chorus begins: 'I want you to hew, little pal, like your dad used to do, little pal. I want you to timber, to corve and to drill, To pack up your canch so's your marra can fill.' The final lines refer to an old bone of contention in miners' households. Because men 'working marras' might share a single paynote, it was not difficult for the miner to retain extra pocket money for himself by pretending his pay was less than it really was, and that his mate had the paynote. In the North-East the practice was called 'keepy back'.

GEORDIE BLACK. Written by the Gateshead comedian Rowland Harrison in 1872. Transcribed from the singing of Peter Elliott, Birtley, Co. Durham (31 July 1963). The 'Gyuss' is the Goose Pit, near Holmside. The 'Betty' is the nickname of a pit near Birtley.

AN OLD MINER. Written by Mike Harding, of Manchester, about an acquaintance, an elderly collier made redundant when the Moston Pit was closed. Mr Harding says: 'I don't know if it's presumptuous to write such a song. I know it's more or less what John feels. Basically, all I've done is taken his words and set them to a tune.' The tune is that of *Come all ye tramps and hawkers.*

THE BANKS OF THE DEE. From James White, Houghton-le-Spring, Co. Durham (7 May 1951). Mr White says: 'It has been sung in the Durham coalfield for many years. As the words imply, it is a relic of the days of coal-owner tyranny, and quite conceivably could be based on fact. . . . To appreciate this song, it is necessary to understand the system of wage-basing in Co. Durham. If men earned more than ten per cent over County Average, the owners could and did apply for a reduction in score, or tonnage rates, and thereby ensured that prices were never high.' One may wonder why a song closely connected with Co. Durham should be set on the banks of the Dee. In fact, a song called *The Banks of the Dee* (whose words, by an Edinburgh judge, John Tait, concerning the departure of a young man to fight in America, bear no relation to our text) was published in 1775, and appeared on a broadside towards the end of the eighteenth century by Angus of Newcastle, and evidently it enjoyed good popularity, for in an abbreviated version it was reprinted four times by J. Marshall, likewise of Newcastle, and an 'Answer to the Banks of the Dee' was also reprinted several times by north-eastern houses. The original publication and the Angus print prescribe the melody: *Langolee*. A fiddle version of this tune appears in Oswald's *Caledonian Pocket Companion*, Book IX, *c.* 1755. The *Langolee* tune fits our coal-mining song well, and perhaps the maker of the song had this melody in mind; if so, the text later acquired Mr White's tune, which is quite different from *Langolee*. Pete Elliott suggests that in the miners' song, the 'Banks of the Dee' may be a punning reference signifying the surface of the 'D' pit.

THE OLD CLOG-DANCER'S SONG. Sung by Pete Elliott of Rickleton, Co. Durham (23 August 1976). He had the song from an old miner, Johnson Ellwood, of Chester-le-Street, Co. Durham, himself once a champion clog-dancer.

EE AYE, AA COULD HEW. Written by Ed Pickford, formerly of Washington, Co. Durham, who ironically styles himself the 'Noel Coward of the Coal-fields'. The song is inspired by the experience of Mr Pickford's father. Transcribed from the singing of Ian Fraser, at Birtley, Co. Durham (6 March 1971).

THE COLLIER'S LAMENT. Text from J. T. Huxtable, of Workington (1951). The song seems to be related to a lengthy broadsheet ballad: *The Submissive Petition of the Distressed Hugh Boyd, Late a Collier, at Parton, near Whitehaven, Cumberland*, containing a prose account of an injury in a pit diaster in May 1784, and an appeal for cast-off clothing, ending: 'If you do not buy this, pray keep clean till called for.'

THE WALLSEND EXPLOSION. The edited version of a text from J. S. Bell, of Whiston, Lancs, who found it as a manuscript in the Picton Library, Liverpool. The explosion occurred on 18 June 1835, at Russell's Colliery, a notoriously gassy pit in the Wallsend district. 102 people were killed, 76 of them under the age of 21 (one victim was but eight years old). At the time of the explosion the hewers had finished work, and most of them had gone home or to the Newcastle Races (it being Gold Cup day). The majority of those left below were the putters, rolley-drivers and trapper boys. 'Mr B' refers to John Biddle (1773–1843), general supervisor of the Wallsend Colliery and an expert in scientific ventilation. He had already warned that the local seam was 'prodigiously fiery'.

THE TRAPPER BOY'S DREAM. From a broadside titled *Dreadful Explosion at*

West Moor Colliery, April 3rd, 1845 in which ten were killed including four boys. The broadside was printed 'at the Miners' Advocate office, Sun Inn, Side' (Newcastle). The explosion at the West Moor Colliery, Killingworth, occurred between two shifts. The gas was ignited by the candle of a little boy who was allowed to proceed before the men. There had been a fatal explosion at West Moor in January of the previous year, but nothing had been done to improve safety. Dreams presaging disaster are not uncommon in the song repertory of pitmen (e.g. *The Miner's Doom*, in which the warning comes to the collier's wife, and the successful music hall song *Don't go down in the mine, dad*, by Robert Donnelly and Will Geddes, published in 1910).

THE JARROW PIT EXPLOSION. From a broadside published by W. S. Crow, of Gateshead, subtitled *Dreadful Explosion at Jarrow Pit on Thursday afternoon, August 21st 1845. Forty lives lost!* The explosion happened about 1 p.m. The Low Main seam where the explosion occurred was worked entirely with naked lights. Seventy-five men and boys were at work there at the time. The number of fatalities was in fact 42.

THE ST HELEN'S COLLIERY EXPLOSION. From a broadside published by J. Harkness, of Preston. In the Mitchell Library, Glasgow. The explosion on 23 June 1849 was at Messrs F. Johnson's Laffak Colliery, Safford, Lancs. It was very violent, but remarkably only eight miners were killed. Though the pitmen were issued with safety lamps, one of the men persisted in working with a naked light, and the explosion was due to his conservatism.

THE FATAL COAL-PIT EXPLOSION AT NITSHILL. From a broadside, no imprint or date, in the Murray Collection, Glasgow University Library. The explosion at Nitshill, Renfrewshire, occurred on 15 March 1851. Of 63 men working in the pit, only two survived.

THE LATE COLLIERY EXPLOSION AT PATRICROFT, WIGAN. From a broadside printed by D. Thomas, of Wallcroft, Lancs. The Patricroft Explosion occurred on 20 February 1854. Eighty-nine men and boys were killed, and 11 injured.

THE CUMMER COLLIERY EXPLOSION. From a broadside, no imprint or date, in the Sheffield University Library's collection. The explosion at the Cummer Colliery, near Pontypridd, happened on 15 July 1856. One hundred and ten miners were killed.

THE HARTLEY CALAMITY. From a poem by Joseph Skipsey (1832–1903). This version from a leaflet. The poem was later printed in Skipsey's *Carols from the Coalfield* in slightly different and expanded form. The disaster occurred on Thursday, 16 January 1862, at Messrs Carr's Hester Pit, when the 40-ton cast-iron beam of the pumping engine broke and a large part of it fell down the colliery's single shaft, bringing with it a great mass of debris. The accident happened just at changeover time, when both shifts – 204 men and boys – were still underground. Their only escape-route was blocked, and all were suffocated. Most of the victims were brought up some ten days later and, laid out in rows, were identified by a 14-year-old boy who ordinarily collected the miners' tallies, and thus was familiar with the looks of each man. A week after the accident, Queen Victoria sent a telegram: 'The Queen is most anxious to hear that there are hopes of saving the poor people in the colliery for whom her heart bleeds.'

THE OAKS PIT EXPLOSION. From a broadside, no imprint, in the Sheffield University Library's collection. The Oaks Pit, near Barnsley, Yorks, blew up on 12 December, 1866, killing 348 miners. It was among the biggest pits in Britain, on one of the new coalfields of nineteenth-century expansion. The Oaks disaster came as the climax to a series of catastrophes in the 'fiery Barnsley bed'. Ploughmen five miles away found the ground covered with a fine sprinkling of coal-dust soon after the accident.

THE STARLAW DISASTER. Collected by Hamish Henderson from Rab Morrison, Woolmet Colliery, Midlothian. The Starlaw accident happened on Saturday, 9 April 1870. It was caused by some timbering in the upcast shaft catching fire. At the time, 56 men and boys were working in the pit, and most of them were rescued, though it meant dragging them through the flames that filled the shaft. At length the cage-rope was burnt through, leaving seven men to die. 'John Steel' referred to in the song was the engineman James Steel, who saved the lives of most of his mates by working the cage though exposed to scorching heat. The song, with many details, was printed in *Scottish Studies*, Vol. 7, Pt. 1, 1963, pp. 92–100.

MELANCHOLY EXPLOSION AT THE MOSS PIT. From a broadside, no imprint or date, in the Sheffield University Library's collection. The Moss Pit at Ince, near Wigan, blew up on 6 September 1871, killing 70 miners. It was resolved to cover in the shafts as the only way to put out the fire. The pit was opened on 21 September when, according to the prose account on the broadside, a 'second most Terrific Explosion took place, with a report like a park of Artillery, and an immense sheet of flame shot up to a great height. Out of the number standing at the Pit's mouth, 5 were instantly killed, and a number seriously injured.'

THE PIT ACCIDENT AT PELSHALL. From a broadside, no imprint or date, in the Sheffield University Library's collection. The Pelshall Hall Colliery in Staffordshire was flooded on 14 November 1872, and 22 miners were drowned.

THE BLANTYRE EXPLOSION (I). Melody and fragmentary text from R. Greening, Glasgow (February 1951). Additional text from Mrs Cosgrove, of Newtongrange, Midlothian (11 May, 1951). The disaster happened at Messrs William Dixon's colliery, High Blantyre, near Glasgow, 22 October 1877. Of 233 men and boys working in the pit at the time, 207 were killed. The management were felt to be much to blame, their economies resulting in poor ventilation and little inspection. Naked lights were used almost throughout the colliery. The miners were aware of the unusual danger but, said one, 'We durst not make any complaint because we would get no redress', and another declared: 'Had I not been long idle previous to going to the colliery, and the rent becoming due, I would not have gone down that pit.' It took the eloquence of the miners' leader, Alexander McDonald, to dissuade the indignant pitmen from violence against the owners and officials. The song has lasted well, and spread to America where Samuel Bayard collected a version from a Pennsylvania singer, Mrs Jennie Craven. *The Blantyre Explosion* is a sprig off the same (ultimately Irish) tree as *Lost Jimmy Whalen*, a persistent ballad in the American North-East.

THE BLANTYRE EXPLOSION (II). A high proportion of the miners killed at High Blantyre were Irish. Perhaps that is why sundry versions of the song commemorating the explosion circulate among Irish workers at home or in emigration. The present example was recorded by Robin Morton from John Maguire, Tonaydrumallard, Co. Fermanagh (August, 1968), and printed in Morton's *Folksongs sung in Ulster* (Cork, 1970). In a private communication Robin Morton says: 'John learned this while working as a miner in the Glasgow area in the 1920s. He tells a lovely story about them all sitting in a pub in Blantyre one night and they sang different versions of the song. This version was naturally decided on as best.' The *Glasgow Herald* (23 October 1877) reported that when the bodies of the burnt and suffocated miners were brought up, there were 'feelings of a painful intensity amongst the at all times excitable Irish element. The women set up a loud wail, tore their hair, and rushed about in a half-crazed state, and strong men had to interfere to prevent them throwing themselves on the corpses.'

FEARFUL COLLIERY EXPLOSION IN SCOTLAND. From a broadside, no imprint or date, in the Mitchell Library, Glasgow. The song refers to the explosion at High Blantyre.

THE TRIMDON GRANGE EXPLOSION. As sung (one verse only) by R. Sewell, of Newcastle (June 1951). Remainder of text from J. Jefferson, Trimdon Grange, Co. Durham. From a ballad by Thomas Armstrong, who prescribed for it the tune of *Go and Leave me if you Wish it*; now it is usually heard attached to the come-all-ye type tune given here. The explosion occurred on 16 February 1882. Seventy-four miners were lost (six of them died in the neighbouring East Hetton colliery, due to afterdamp seeping through from Trimdon).

TERRIBLE COLLIERY EXPLOSION AT CLIFTON HILL PARK. From a broadside, no imprint or date, among the Kidson Collection in the Mitchell Library, Glasgow. The accident was at Messrs William Knowles's colliery at Pendlebury, near Manchester, 8 June 1885. One hundred and seventy-eight miners were killed.

THE DONIBRISTLE MOSSBURN DISASTER. As sung by J. Ferguson, of Markinch, Fife (May 1951). The accident at Donibristle, Fife, occurred on 26 August 1901. Much of the mine-workings lay under Moss Morran, a stretch of soggy moorland. The seam was worked on a steep incline up towards the surface, the surface gave way, and there was an inrush of moss into the mine. Eight men, including the overman Rattray and three men of a rescue party, lost their lives.

Pit disasters due to the inflow of moss are fortunately rare, though the terrible events at the Knockshinnoch Castle Colliery on 7 September 1950, when thousands of tons of liquid peat and moss rushed in where the workings had been driven too close to the surface, are vivid in the memory of many Scottish miners still working underground.

THE MAYPOLE DISASTER. Collected by Dave Hurst, Hindley Green, Wigan, from Mrs Maggie Davies, of Orrel Post, Wigan. The disaster occurred on 18 August 1908, when 75 miners lost their lives. Another version, rather shorter, was obtained from N. McEwan of Dover (formerly of St Helen's), who said: 'I heard that song when I first went into the pit in 1916.'

THE MAYPOLE PIT DISASTER. Another song about the same explosion, collected by Dave Hurst from an 86-year-old ex-miner, Peter Parr, of New Springs, Wigan (1968).

THE GRESFORD DISASTER. Text and tune communicated by Ewan MacColl, 'from the singing of a young miner named Ford in Sheffield Miners' Training Centre'. The explosion and subsequent fire happened on 22 September 1934. The ballad slightly underestimates the number of casualties: 265 miners were killed, including three rescue men. The subsequent Court of Inquiry drew attention to 'innumerable breaches of the law and of regulations laid down for the protection of those who worked in the mine, and the behaviour of the inspectorate had been deplorable'. In the House of Commons, D. R. Grenfell described part of the pit where 'men were working almost stark naked, their clogs with holes bored through the bottom to let the sweat run out, with a hundred shots a day fired on a face less than two hundred yards wide. The air was thick with fumes and dust from the blasting and . . . a space two hundred yards long and hundred wide above the wind road was full of inflammable gas and impenetrable for that reason.' The Inquiry showed that the owners were urging output at all cost, and putting increased pressure on the men, and 'in an area like that, where a miner's only hope of employment was at the Gresford pit, it was useless to suggest that the men could make themselves offensive to the management by complaining of circumstances that were perfectly well known to the management before the complaints were made'. For a number of charges of negligence, the colliery company was fined some £15.

THE EASINGTON EXPLOSION. The mine blew up on 29 May 1951, causing 81 deaths (including two rescue men killed while the bodies were being recovered). The song is by George Purdon, former pit deputy, of South Pelaw, Chester-le-Street. He set it to the tune of *Come all ye tramps and hawkers*.

THE AUCHENGEICH DISASTER. The disaster, caused by an underground fire, occurred on 18 September 1959. The death roll was 47. The version here is as sung by Dick Gaughan on a Topic record (12TS 271). He learnt it from Geordie Hamilton of Edinburgh, 'who was brought up in the Lanarkshire coalfield'. The tune is a variant of the well-known *Skippin' barfit thro' the heather*.

THE WATTER'S BROKKEN IN. A song by Bert Draycott, of Fishburn, Stockton-on-Tees, transcribed from his singing. At the Fishburn Colliery in December 1968, an inrush of water occurred, due to working too close to the limestone layer above the coal measure. The danger had been pointed out by the miners and the colliery management, but the local officials of the National Coal Board opted for the risk, since the seam was a productive one, and there was much pressure for increased output. 'Some pin-striped arses needed kicking', said a pitman, commenting at the time; and that is the gist of the song's last verse. Greenlanes was the National Coal Board's headquarters for the district. An Imeco is a small electric bulldozer used to shovel the muck away within the developments.

JOHNNY SEDDON. From the singing of Thomas Mitchell, Chopwell, Co. Durham (November 1953). Mr Mitchell had the song from his father, an Antrim man.

THE MINER BOY. This version of the above ballad, sung by John Doherty, Glenties, Co. Donegal, is in the BBC Folk Song Archive (No. 19588, recorded by Peter Kennedy and Sean O'Boyle in August 1953). In Doherty's version the lost miner is named John Sneddon, as on the broadside copies.

THE LOVER'S LAMENT FOR JOHN SNEDDON, A COLLIER. From a broadside, no imprint or date, in the Sheffield University Library collection. On the broadside, the text is attributed to 'John Wilson, B.S.'.

THE MINER'S DOOM. Version edited from texts communicated in 1951 by E. D. Williams, of Brymbo, Denbigh, and from James Hedley, of Aberavon, Glamorgan. Cf. George Korson, *Minstrels of the Mine Patch* (Philadelphia, 1938), p. 203, where the prescribed tune is given as *Adieu to dear Cambria.*

HEROES, BRITISH HEROES. One of the most widespread of miners' songs. The version here is from the singing of Mrs J. Pattison, of Clowne, near Chesterfield (June 1951). A similar version, less complete, was recorded about the same time from F. Woodvine, of Siverdale, Staffs., who said: 'I first heard it at Madeley Collieries, Leycett, North Staffs., some twenty years ago.'

COLLIER HEROES OF THE MINE. Text from Sam Arkwright, of Trimdon Station, Co. Durham (July 1951), who said: 'My parents used to sing it above fifty years ago. I can't tell if it was about any special disaster. We were living in Lancashire then.'

RAP HER TE BANK. Collected by W. Toyn from Henry Nattress of Low Fell, Gateshead (1962). The song has enjoyed enormous popularity within the last decade or so, flowing probably from the influential folk song club at Birtley, Co. Durham.

THE MAN CALLED LEN. From Barbara Allen, aged 13, who learnt it from her father, a miner at Denaby Main Colliery, near Doncaster (8 February 1964).

THE PITMAN'S HAPPY TIMES. Words by J. P. Robson, 'Bard of the Tyne and Minstrel of the Wear' (1808–1870), from *Bards of the Tyne* (ed. J. P. Robson, 1849). Directed to be sung to the tune of *In the days when we went gypsying.* During the first half of the nineteenth century, miners' wages and conditions differed greatly from coalfield to coalfield, pit to pit, season to season. In some areas, poverty was consistent, in others, the miners were among the better-paid workers. In Durham in the 1840s, a time of trade depression and labour troubles, there was a general decline in living standards, reflected in this song.

THE MINERS' BINDING. From a broadside published (1822?) by Pollock, of North Shields. Miners in Northumberland, Durham and some other parts of the country were employed on a yearly bond, whose terms included rates of pay, conditions of employment, scale of fines for unsatisfactory work. The bond's feudal origins, dating back to the sixteenth century, were clear. Its effect was to tether the miner to his work and to thwart his attempts at betterment. One of the first great strikes on the Tyne and Wear (1810) was a result of dispute over bonding. The strike lasted seven weeks, the military were called in, the gaols were filled, and several pioneers of later union struggles, notably Ben Embleton, received their 'baptism of fire' during those wild weeks.

THE FIRST DREST MAN OF SEGHILL. From a broadside printed by J. Marshall of Newcastle (n.d., but a handwritten inscription – John Bell's? – on the Sheffield University Library copy indicates '31 March 1831'), with the engraving of a man tossed in a blanket. Communicated by J. S. Bell, of Whiston, Lancs. The metre is that of a sizeable family of Irish songs whose members include the well-known *Bansha Peeler* and *The Peevish Child* (Petrie, No. 151), also the envangelical hymn *Babe of Bethelehem*, known to many Primitive Methodist congregations ('Ye nations all, on you I call, / Come hear this declaration, / And don't refuse the glorious news / Of Jesus and salvation'). Any of these may have been in the ballad-maker's head as he wrote the song.

THE PITMEN'S COMPLAINT. From a broadside 'printed for the author by W. Fordyce, 48 Dean Street, Newcastle'. The copy was formerly among the papers of John Bell, and bears a pencilled note 'May 13th 1831'.

THE PITMEN'S STICK. From a broadside printed by Douglas and Kent, 50 Quayside, Newcastle. Subtitled 'A new song by Jeremiah Knox'. Likewise from the papers of John Bell, with a pencilled inscription on the leaflet: 'July 1831'.

THE BLACKLEG SONG. Text contributed by Thomas Holmes, Fence Houses, Co. Durham (1951). From a manuscript scrapbook of the period *c.* 1831.

THE PITMEN'S UNION, or THE LADS OF WEAR AND TYNE. From a broadside printed by Stephenson of Gateshead, among John Bell's papers, with the note 'July 1831'. Conventional as such pieces are, they offer clues about the time. Thomas Hepburn, of Hetton Colliery, Co. Durham, was the leader of a small union. On 21 March 1831, some 20,000 miners met on Newcastle Town Moor to protest against insecure employment, the truck system, and the fines allowed by the bond. Their demands resulted in a lock-out, beginning on 5 April, which lasted for two months. Eventually the men gained several concessions, and the song reflects their mood of victory. However, the government was alarmed, steps were taken to break the union, Hepburn was victimized and, unable to find work, he wandered round the pit villages for some years, selling tea from door to door. *The Pitmen's Union* has been described as 'the battle-hymn of the Hepburn union'. In Spring 1976, the GPO issued a stamp commemorating Tommy Hepburn as a 'Social Reformer'. How surprised he would have been!

THE OPPRESSIONS OF THE PITMEN. From a broadside printed by Douglas and Kent, Newcastle, 'composed by Joseph Hall, collier, Sheriff Hill'. The song was apparently made in late March or early April 1831, between the time of the big meeting on the Town Moor and the actual lock-out.

THE PITMEN'S AGREEMENT. A broadside printed by Douglas and Kent, Newcastle-upon-Tyne. Subtitled 'Written in Commemoration of a great Number of them gaining their Rights, after a very Long Contention with the Coal-Owners, in the year 1831'.

THE TOOLS OF SHOTTON MOOR. 'A New Song by William Hornsby'. From a collection of (mainly) manuscript papers in the Picton Library, Liverpool, relating to events in the Durham and Northumberland coalfields, 1832–44. The metre, not common in England, is more familiar in Scotland – e.g. *Glasgow Peggy* (Child 228) – and particularly in Ireland, where one of the most widespread examples is *Tow Row Row*, or *Paddy, Will Ye Now*. Probably Hornsby had some such model in mind.

THE RECHABITE MAN. From a broadside, no imprint, in the Picton Library collection. The song is undated but it appears to relate to 1835. The author is given as John Henderson. The Rechabites were members of a temperance body, quite powerful in the coalfields during the years when Primitive Methodism was in the ascendancy among miners.

COME ALL YOU CANNY PITMEN. From a broadside published (No. 148) by George Walker, Jun., of Durham. Printed along with *The Wind Blew the Bonny Lassie's Plaidy Awa*. The print is undated, but was evidently issued in 1836.

MR ROBERTS THE PITMEN'S FRIEND. From a broadside, no imprint, undated (but apparently 1844), printed along with *The Miner's Friend*, a song written by W. Johnson of Framwellgate Moor. Prescribed to be sung to the tune of *William and Phyllis*. Thornley Colliery, near Tow Law, Co. Durham, was one of the first to be unionized, and it long remained a stronghold of the Miners' Association. Chafing against the bond that allowed viewers to impose heavy fines, the Thornley men struck in 1843 when 47 of them were fined because their coal contained stones. Warrants were issued against strikers for breaking their bond, and the first cases were heard on 7 December 1843. Due to the eloquence and pertinacity of the miners' advocate, William Roberts, the cases were dismissed and Roberts, returning, was carried shoulder-high for four miles into Thornley. We give the *William and Phyllis* tune as collected by R. Vaughan Williams from three men in the streets of Westminster in 1904, and printed in the *Journal of the Folk-Song Society*, No. 8, p. 216.

A DIALOGUE BETWEEN PETER FEARLESS AND DICK FREEMAN. A ballad of the

1844 Durham strike. From a manuscript copy among papers relating to events in the north-eastern coalfield in the 1830s and 1840s, in the Picton Library, Liverpool. Communicated by J. S. Bell, of Whiston, Lancs.

THE MINERS' GRIEVANCES. From a broadside, no imprint, undated (but evidently 1844), in the Picton Library collection. The author is given as William Johnson of Framwellgate Moor. Another version, with some alterations, is among a manuscript collection in the Wigan Public Library. From a modest but enthusiastic beginning in Yorkshire in the summer of 1842, the Miners' Association grew fast. Within a year, it had 50,000 members and 14 lecturers (paid agitators, in the official view). At the start of the great Durham strike in April 1844 (Johnson's ballad sets out the issues) the union was about 80,000 strong, with 47 lecturers.

MARTIN JUDE. A modern ballad by George Purdon, of South Pelaw, Co. Durham. Martin Jude (1805–60), treasurer of the Miners' Association till 1846, learned his trade unionism during the strike of 1831–2. Blacklisted by the coal-owners, he became a publican in Newcastle. One of his taverns, the Three Tuns, was often used for union meetings. He was one of the most devoted pioneers of the labour movement. During the 1844 strike he handed over his life savings to the union. It seems the debt was never repaid, and he died in deep poverty.

THE BALLAD OF ROBERTS. Text communicated by J. S. Bell, from the Picton Library collection. William Prowting Roberts (1806–71) was a solicitor from Bath, son of a wealthy Cheltenham vicar, nephew of Sir Nicholas Tindell, the Lord Chief Justice. A confirmed Chartist, Roberts was sentenced to two years' imprisonment in 1839 for his political activities. Prison experience and Chartist contacts brought him to the miners, and he was engaged by the Durham and Northumberland union in 1843 to be their attorney and to fight every case against the pitmen in the law courts. Through eloquence and doggedness, he won case after case, and within a year of his appointment he was the miners' hero. At times, his devotion meant severe sacrifices and financial distress. After the failure of the 1844 strike, the *Gateshead Observer* wrote: 'Roberts is still lurking about the district, so miserably clad and dirty in person, he might be taken for a hewer in his second best.' Of himself, Roberts wrote: 'There is no labour so grateful as that of opening prison doors and letting the oppressed go free.'

COME ALL YOU COLLIERS. From a manuscript in the Picton Library collection. Communicated by J. S. Bell.

THE SONG OF THE 1844 STRIKE. From a broadside, no imprint or date. Author marked as 'Anon'. Prescribed air: *Auld Lang Syne.*

THE DURHAM STRIKE. From a broadside, no imprint or date, in the Newcastle-upon-Tyne Public Reference Library. The author's name is given as W. Roxby.

THE COAL-OWNER AND THE PITMAN'S WIFE. Text communicated by J. S. Bell, from a manuscript in the Picton Library Collection. The author is thought to be William Hornsby, who also wrote *The Tools of Shotton Moor*. (Incidentally, if, as the song suggests, the writer was evicted on 13 May 1844, he must have been among the first to be turned out of his home by the coalowners.) The tune and a fragment of text was got from J. Denison, of Walker, Co. Durham, in June 1951. The song has obtained fairly wide currency in the edited version published in the first edition of the present work (1952).

THE 1844 STRIKE. From a broadside, no imprint or date. Directed to be sung to the tune of *The Grand Conversation under the Rose*. Evidently the song was made during June 1844, when the strike was already ten weeks old.

THE PITMEN'S UNION (I). From a broadside, no imprint or date, in the Newcastle-upon-Tyne Public Reference Library. The authors are named as William Walker Story and George Watson, of West Rainton. The song is prescribed to be sung to the tune of *And a-nutting we will go*. The 'noble Marquis' was Lord Londonderry, Lord Lieutenant of Durham, half-brother to Lord Castlereagh, and one of the most powerful coalowners in the North-East. He proceeded against the strikers ruthlessly, with arrests, mass evictions (those at West Rainton took place at the end of June), the large-scale importation of blackleg labour from his estates in Ireland, and the victimization of shopkeepers and tradesmen who gave credit to striking pitmen.

THE PITMEN'S UNION (II). From a broadside, no imprint, dated 1844, in the Newcastle-upon-Tyne Public Reference Library. The pious tone of the final verse need not surprise; more surprising is that such a tone is not commoner in the strike songs. Particularly the Primitive Methodists, or Ranters, were numerous among the early trade unionists, and often lay preachers, being men with a gift for eloquence and argument, were chosen as spokesmen to present the men's grievances to the owners. Especially during the 1844 strike, union meetings were often held in Ranter chapels, and prayers publicly offered up for a successful outcome of the strike.

THE PITMEN'S GRIEVANCES. From a broadside, no imprint or date, in the Newcastle-upon-Tyne Public Reference Library. Particular interest attaches to this song, as to *The Colliery Union*, since these are among the few that we know were made by women. In the present case, the author is named on the leaflet as Jane Knight. Her song was probably made in July 1844, for by then whole 'villages' of ramshackle shelters had sprung up on the moors as a result of the evictions.

THE COLLIERY UNION. From a broadside printed by Henderson of North Shields. The author proudly signs her work: 'Elizabeth Gair, Collier's Wife'. The song was probably made about the same time as *The Pitmen's Grievances*, for it was not until June 1844 that the owners began to bring in blacklegs from Wales, Cornwall, Scotland and Ireland. William Beesley was an Accrington chainmaker who became a Chartist journalist, editing the *North Lancashire and Teetotal Letter-Bag*. He served as clerk to W. P. Roberts and became editor of the first miners' newspaper, *The Miners' Journal*.

THE BLACKLEG MINERS. As sung by W. Sampey, of Bishop Auckland, Co. Durham, 18 November 1949. George Korson, in *Coal Dust on the Fiddle* (Philadelphia, 1943) prints what looks like a parody of this song, called *The Yahie Miners*. Korson's version comes from Glace Bay, Nova Scotia, a well-known changing-post for British and American miners' songs. The Durham song has become quite widespread since its appearance in the first edition of the present work, and the tune in particular has taken on variant shapes.

THE TRUCK MASTERS. From a broadside, no imprint or date, in the Murray Collection, Glasgow University. The truck or 'tommy' system was originally payment in goods instead of cash. Later, the meaning became modified to imply that the employers acted as provision merchants, supplying their workmen with food, soap, candles and such. In the description of the Midland Mining Commission (1843): 'One of the essentials of the tommy system is to pay the wages only once a month or five, six or seven weeks. Now, as the men cannot go without any fresh supplies for so long a period, their only recourse is to apply to the masters' shops and to get goods in lieu of part of what they would otherwise receive from the butty at the reckoning.' One objection was that the goods were inferior (hence the expression 'tommy rot'); also prices were higher than in ordinary shops; finally, a collier in debt to the master would be in a poor position for bargaining over wages and conditions. Trade union action did much to diminish the truck system in England by the 1860s, but in Scotland it lingered on for at least a couple of decades more. The present ballad was evidently made some time between 1841 and 1859. Miller and Eadie were joint owners of the New Monkland Colliery at Airdrie, and they established the Langloan Iron Works in 1841. They were notorious truckmasters, as were the Wilsons of Summerlea. The Gartsherrie Ironworks at Coatbridge abandoned the truck system in 1859. William Dixon, whose family had been important in Scottish heavy industry since the end of the eighteenth century, owned at least five collieries in the Glasgow, Lanarkshire and West Lothian areas. He was considered a model employer, who paid good wages and did not operate the truck system.

THE COLLIER LAD'S LAMENT. From a broadside printed by H. Pratt of Birmingham, undated, but evidently written during Lord John Russell's prime-ministership (1846–52). Robert Galloway, in his *Annals of Coal Mining* (Second Series, London, 1904), remarks: 'The extraordinary extent to which the Truck Act was evaded at this period had the effect of almost reducing some of the coal and iron districts to the primitive conditions of communities devoid of metallic currency, where payments had to be made in kind, the great bulk of the wages consisting of notes on the "tommy shop"'.

CALLERFORNEY. Words by J. P. Robson (1808–1870), from *Songs of the Bards of the Tyne* (J. P. Robson, ed., W. S. Crow, Newcastle, 1849). Directed to the sung to the tune of *Alley Croaker*. Robson must have been quite quick off the mark, for though the great discovery of Californian gold was made on 24 January 1848, the news hung fire for some time and it was not till the following year that the great rush of gold-seekers, the celebrated 'Forty-niners', got under way, amid a turmoil that lasted till 1856. At latest, Robson must have made his song at the very start of 1849, just before the book was printed.

ASSTRILLY, or THE PITMAN'S FAREWELL. Words by the stage comedian Edward Corvan (1830–65). Directed to be sung to the tune of *All round my hat*. Written some time during the great gold rush period in Australia which lasted through the half decade from 1851 (though in fact, particularly in Queensland, the goldfields were attracting a fair flow of immigrants till the late 1860s). *Asstrilly* was among Ned Corvan's successes when he appeared with the Olympic Company in the music hall on the site of Madame Tournaire's circus on the Newcastle public grounds, the Forth, early in the 1850s.

BANG THE BLEAZER, BEN. A modern song by George Purdon, of South Pelaw, Co. Durham. To the tune of *Bringing in the Sheaves*. Ben Embleton, of Wingate Colliery, Co. Durham, was a veteran of the miners' struggles who had participated in strikes as early as 1810. With Tommy Hepburn, he was one of the leaders of the movement to form a union in 1831, and was prominent during the bitter years of the 1840s. As a result he was cruelly victimized by the coal-owners. During the upsurge of union activity in 1849, according to R. Fynes in his *Miners of Northumberland and Durham* (1873), 'it was difficult to get meetings in many places, and old Ben Embleton was to be seen on many occasions going about the collieries with a tin, or a sheet of iron, commonly called a "bleazer", to attract the attention of the men and induce them to come out.' Mr Purdon's song is doubtless based on Fynes's account.

THE COLLIERS' EIGHT HOUR DAY. From a broadside printed by Robert McIntosh, Gallowgate, Glasgow. It is undated, but probably belongs to the period shortly after the events of June 1870. Another broadside, without imprint, having almost identical text, gives John Wilson, collier, as the author. Alexander McDonald (1821–81) was born in Lanarkshire and entered the mines at the age of eight, working fourteen hours a day. After the 1842 strike, he was victimized and obliged to leave the pits. By this time he was 22, and remarkably enough for a self-educated young man of the period, he was able to study for a while at Glasgow University. Just as remarkably, he became a mine manager but still spent much time agitating on the miners' behalf. In 1870 he campaigned throughout the industrial areas of Scotland, in particular on behalf of the Eight Hours Movement. In June 1870, following a stay-down strike, the Fife pitmen secured an eight-hour day, the first miners in Europe to do so. In 1874, McDonald became M.P. for North Staffordshire, and chairman of the Parliamentary Committee of the TUC. Towards the end of his life, he moderated his views somewhat, and though the miners mourned him when he died, there were some who thought him an adventurer who 'used the miners to help him win a great place in the world'. The present ballad, set to the tune of *The Bonny Lad that Handles the Plough*, is to be heard on a Topic record (12TS 226).

THE MINERS' CRY. From a broadside, undated, printed by James Lindsay of Glasgow, in the Murray Collection, Glasgow University. The ballad is printed along with the text of the Napoleonic song *The Bunch of Roses O*. The 'grand news' referred to in the first line of the ballad was doubtless the passing of the Mines Act of 1872, a triumph for Alexander McDonald after his fight for higher wages, shorter hours, better safety conditions for miners. The Scottish miners presented him with £680 and a silver casket 'to keep his siller in'.

THE SOUTH MEDOMSLEY STRIKE. Text from John H. Denny, of Stainforth, Doncaster (December 1950), formerly of Tantobie, Co. Durham. The words are by Thomas Armstrong, and Mr Denny's version doubtless derives from the *Song Book Containing 25 Popular Songs of the late Thomas Armstrong, Compiled by his son W. H. Armstrong* (Chester-le-Street, 1930), where the text appears in a rather knotted orthography imitating the local 'pitmatic' dialect. The prescribed tune is *Castles in the Air*. At South Medomsley, near Consett, Co. Durham, the 1885 strike was remarkable for the number and roughness of the evictions of strikers from their company-owned homes. Strictly, candymen were rag, bone and scrap metal merchants who rewarded children with sweets in return for waste material. In this context the term refers to the lumpen-proletarian idlers who were enlisted to help in the evictions, under the protection of the police. A performance of the ballad is to be heard on a Topic record (12TS 271/2).

THE OAKEY STRIKE EVICTIONS. Words by Thomas Armstrong. Originally directed to be sung to the tune of *The Pride of Petticoat Lane*, but over the last twenty years or so it has been quite widely current under the melody given here. The events occurred during the November 1885 stoppage in the Stanley–Consett–Chopwell area. Interviewed in August 1951, Tommy Armstrong's son William 'Poety' Armstrong recounted the origin of the song. Tommy was jealous of his reputation as a local bard. A miner from Gateshead, William Maguire, got work in Tanfield and boasted of his skill as a song-maker. A 'duel' was arranged between the rivals, the miners gathered in the Red Roe public house, and lots were drawn as to who should take first turn in improvising a song on the subject chosen by the company. The theme of The Miner's Lamp was rejected, likewise that of A Pint of Ale. Finally, the topic of the evictions was approved. Maguire's song is quite forgotten, but Armstrong's (subsequently 'brushed up' by him at leisure) remains, and indeed has taken a new lease of life since 1952.

THE MINERS' SONG OF FREEDOM. From a broadside, no imprint or date, among papers relating to the Somerset Miners' Association (SMA) in the Wells Memorial Library, Bristol University. Communicated by Tom Randall of Midsomer Norton, Somerset. The song is marked 'As sung by the Somersetshire Miners on strike, 1889', and signed 'composed by An advocate of Right against Might'. The prescribed tune is *Onward, Christian Soldiers*. Mr Randall suggests that the author may be G. Mitchard, an SMA official. Samuel Henry Whitehouse (1849–1919) referred to in the song, was a

Staffordshire man who entered the pits at an early age. He was made secretary of the Midland Counties' Federation of Miners in 1886, but two years later he was invited by the Somerset miners to come and organize them. He remained their secretary for thirty years.

THE DURHAM LOCK-OUT. Words by Thomas Armstrong. Originally directed to be sung to the tune of *Castles in the Air*, but nowadays current under the tune printed here. The lock-out began on 18 March 1892, when the Durham owners proposed a ten per cent wage-cut on the ground of a fall in coal prices. The miners rejected the proposal and were locked out. After six weeks, with the pitmen's families hard hit and begging for food, the owners proposed to allow the men to resume work with a wage reduction of thirteen and a half per cent. Armstrong's ballad must have been written just at this point. The men refused, the lock-out continued; eventually, after twelve weeks, the owners consented to reopen the pits and the men accepted the ten per cent reduction.

THE MINERS' LOCK-OUT. From a broadside, no imprint or date, among papers in the Pitmen's Strike Collection in the Wigan Public Library. The broadside is embellished with a cut showing a colliery, also two clasped hands with the motto: 'United we Stand! Divided we fall!' The song is marked: 'Written and Composed by Burnett O'Brien, Wigan', and the specified tune is *Castles in the Air*. The piece dates from 1893. Following a fall in the selling price of coal, the owners demanded a twenty-five per cent reduction in wages. The miners refused to accept the demand, and in July the owners declared a lock-out, affecting some 300,000 workers in the area covered by the Miners' Federation. The lock-out lasted sixteen weeks, marked at times by violent disturbances. For instance, on 7 September at Featherstone, Yorks, a riot ensued when soldiers were brought to the colliery, the crowd was fired on, two men were killed and sixteen wounded. Eventually, on 17 November the settlement was signed, the miners were victorious and returned to work at their old wage. Samuel Woods (1846–1915) of Wigan was the first vice-president of the Miners' Federation of Great Britain, elected in 1889.

WHO WILL HELP THEM?. From a broadside, no imprint or date, among the Somerset Miners' Association papers in Bristol University Library. The song is signed 'G. Mitchard', and the prescribed tune is *He'll be back by and by*. Communicated by Tom Randall who identifies the song as relating to the strike of the miners of Radstock, between Bath and Wells, in 1904. The offer of help from the northern coalfields was realized mainly in contributions to strike pay (at that time, non-union men also received strike pay, and the proportion of non-union to union men in Somerset was high).

THE DUNKERTON CARTING BOYS. Collected by Tom Randall from Mrs Lily Parfitt of Radstock, and Mr Charles Nuth of Peasedown St John, Somerset, and collated from their versions. The tune is *I'm a lassie from Lancashire*. In *Bristol Folk News*, No. 15 (Summer 1974), Mr Randall gives details of the song's background. The Dunkerton Colliery was opened in 1906 and closed in

1927. From the first, it was notorious for hard and unsafe working conditions. In 1907 the carting boys, numbering some two hundred, applied for a rise in pay, on the grounds that the putts were heavier and the roads rougher than in other collieries. The current rate was 1d a ton for the first 50 yards, and $\frac{1}{4}$d for each subsequent 50 yards. The boys wanted 1d throughout. The dispute simmered till October 1908, when the boys struck, and the men came out with them. In January 1909, angered over the employment of blacklegs, a threatening crowd gathered and was fired on by Mr Edgar Heal, the manager's son. Ten miners were gaoled and Mr Heal was acquitted of the charge of malicious wounding. The dispute went to arbitration, but the carting boys got no increase.

THE DURHAM MINERS' BIG MEETING. From a broadside, no imprint, dated 18 August 1910. It was a year of high elation for Durham miners. Their union was the second largest in the country (120,000 members) and had relatively recently – 1908 – joined with all the unions from other coalfields into the Miners' Federation of Great Britain, a triumph of the rank and file over a stubbornly obstructive leadership. Hence the excited roll-call declaimed in this song.

MINER'S LIFE IS LIKE A SAILOR'S. Text from George Evans of Aberaman (1951), 'sung to the Welsh hymn tune *Calon Lan*', or more precisely, to the adaptation of that tune used for the evangelical hymn *Life's Railway to Heaven*. Mr Evans remarks, 'Not many miners in South Wales know this one.' It is well-known in USA, however (see George Korson, *Coal Dust on the Fiddle*, p. 413). It was issued on a broadside 'produced for the National Union of Miners, Betteshanger Colliery, May 1958'.

THE CLYDACH VALE LOCK-OUT. From a leaflet marked 'Composed by one of the Unemployed at Clydach Vale'. The prescribed tune is *Just before the battle, mother*. Early in April 1921, following a threat of drastic wage-cuts by the coalowners, the miners issued a strike call and appealed to railwaymen and transport workers to come out in sympathy. In alarm, the Government sent troops to the main industrial areas, and machine guns were posted at the pit-heads. However, some of the trade union leaders, notably J. H. Thomas and Ernest Bevin, threw in their hand and declared ('Black Friday', 15 April) that the strike was off. The workers' united front collapsed, and eventually the miners resumed work on the owners' terms. The employers' attack was mercilessly pressed home, resulting in a wave of local strikes and lock-outs. The Clydach Vale lock-out began in August 1921, and some of the Clydach Vale pitmen were among the earliest of the 'gutter-choirs' of unemployed Welsh miners, trailing along the kerbside singing hymns, a terribly familiar sight in the ensuing years.

FAREWELL TO COLLIERY OWNERS. A song by William ('Poety') Armstrong, of Tanfield Lea, Co. Durham, to celebrate the end of private ownership of the pits by the Coal Industry Nationalisation Act of July 1946. The prescribed tune is *A Good New Year to One an' A'*.

THE POUND A WEEK RISE. A song by Ed Pickford, formerly of Washington, Co. Durham, written in 1963. The BBC considered the song libellous to Lord Robens, then head of the National Coal Board, and banned it from being broadcast. However, though the issue it describes is dead, the song is still (1978) sung with relish in the North-East. Here, text and tune are transcribed from the singing of Ellis Holliday, of Stockton-on-Tees (19 August 1963).

IT'S MINER THIS AND MINER THAT. Written by George Purdon who sings it to a version of *The Lambton Worm* tune. 'Lord Shanky' is in fact Mr Justice Sankey, chairman of the Coal Industry Commission of 1919.

CHEER UP, LADS. From the singing of Bert Draycott, Fishburn, Stockton-on-Tees, who wrote this ironic anti-bureaucratic piece in the late 1960s. Mr Draycott explains the song in some detail: 'At the time, I was in charge of the developments at the colliery, and we bumped solid stone, drilled up, drilled down, and we couldn't find any coal anywhere, and we bored from the surface and discovered that the coal seam, the Harvey seam that's at our colliery, had just petered out and disappeared; there was no coal left. So they decided that the union and the management give us a whole lot of guff that the colliery could carry on with the Busty seam alone, which had shown fantastic loss for the past five years. At the time the Busty seam had one face going that was three foot six high. Out of this, for about seventy yards of a two hundred yard face, there was two foot of really hard post stone in the middle. Well, the trepanner used to go out and cut this face, it used to take him about three days to get through it, where he should be going through five or six times a shift. And they used to change the picks when they were going through this hard stone – the picks cost thirty shillings each, by the way – they used to change the picks about every yard. The lads all used to carry a box of picks. And the big wheel the picks were mounted in, they'll last for evermore under normal conditions, but they used to drop to bits, the weld would break, they'd fall off about twice, three times a week. These cost about three or four hundred pound apiece. As for the bit about the damp in the trunk-gate, well, the deputy could pay the men 1/6 a shift extra for working in what he thought were wet conditions, but the management couldn't trust the deputy to decide what was excessive wet, and the 'howerman' (overman) had to book this in; and some of these howermen thought they had to pay it out of their own pocket. Anyway, the management and the union had a meeting, and they said, oh, they'd make a profit out of the Busty seam. And they had this meeting, and explained it all to us, and I wrote this song shortly after. It was all a load of guff, because you can't make a profit out of solid stone. The idea, I think, was that they rendered it down and got olive oil out of the stone, but it never materialized.'

FAREWELL TO THE MONTY. Written by John Pandrich ('Johnny Handle') in 1957, when the pit he was working in – the Montagu Colliery at Walbottle, near Newcastle – was being closed down, though it seemed there was plenty of workable coal still there. The author remarks that 'the Monty Pit had survived explosions, loss of coal due to geological conditions, strikes and lock-outs'. The song well conveys the ambivalent feeling of miners who hate to work in such a rigorous pit, yet cannot bear to leave it.

LAMENT FOR ALBERT. A song by Keith Roberts of Ashton-in-Makerfield, Wigan, Lancs. Words from a broadside (*North Countrie Broadsheet Balladrie*, No. 5) published by Mr Roberts. The broadside has the legend: 'December 1965 saw the closing of one of Lancashire's friendliest pits – the Albert Pit at Abram . . . only a small pit by most standards but it had retained much of the atmosphere of pre-Nationalisation days. . . . Now alas, owd Albert's gone!!!' The tune is transcribed from the singing of Keith Roberts on a Trailer record, LER 3031.

FAREWELL TO 'COTIA. Written by George Purdon who set it to the tune of *Come all ye tramps and hawkers*. At the time (*c.* 1960) Mr Purdon was deputy at the Harraton pit, called 'Nova Scotia', or ''Cotia' for short, on account of the number of runaway miners from Scotland, fleeing from harsh conditions, who settled in the district from the seventeenth century onward. After more than three hundred years of operation, the pit was gradually 'run down' by the National Coal Board. In 1954 some 1400 men worked there, but within a decade the number had been reduced to less than 500. Some of the men were transferred to Nottingham ('Robens' Promised Land') or South Wales; others left the industry; the older men were made redundant. The song came to have almost ritual significance, being sung at the end of farewell parties of miners under transfer, much as *Auld Lang Syne* is sung in other circumstances.

THE 'COTIA BANNER. A song on the same theme as *Farewell to 'Cotia*, likewise by George Purdon, who set the words to a rather individual version of Henry C. Work's tune, *Marching through Georgia*.

DURHAM BIG MEETING DAY. John Pandrich ('Johnny Handle') made this song, words and music, in 1961. Of the Durham Miners' Gala, which the piece celebrates, he wrote: 'A yearly get-together to discuss grievances, for friends to meet; the inevitable bands and beautiful lodge banners, sometimes even today draped in black to denote the passing of a fellow-worker at the pit in a fatal accident. This day has much more significance than any outing for pleasure or scenery – this is the contact of miners, beer, bands, political speeches, always with the knowledge of the heritage – the reason for its continuance.' A performance of the song, by Johnny Handle, may be heard on a Topic record, 12TS 270.

A MINER'S LIFE. Written by Ed Pickford, *c.* 1972, at a time when the National Coal Board was closing many pits, and the coal-mining industry was being written off as obsolescent. The song won the North East Folk Federation's annual songwriting competition (1973). A performance of it by Tom Gilfellon may be heard on a Topic record, 12TS 271.

THE PITMEN ARE NOT BONNY LADS. From a collection of local songs mainly assembled by John Bell about 1830, now deposited in the University Library, Newcastle-upon-Tyne. Directed to be sung to the air of *Corn Rigs*.

TIN MINERS' SONG. Recorded by Peter Kennedy from the Skinner's Bottom Glee Singers, Redruth, Cornwall, in 1956. The recording is No. 23654 in the BBC's Archive of Recorded Folk Music. Sung to a variant of the tune of *Trelawney* (or, for that matter, *The Brave Old Duke of York*). 'Knackers', or 'Tommy-knockers' are kobold-like spirits of the mine, sometimes benevolent, sometimes not. 'Bucca' is a familiar name for such a spirit.

THE DRUNKEN EXCISEMAN. From *The Scottish Comic Songster* (n.d., *c.* 1850), a twopenny song book issued by John S. Marr, 194 Buchanan Street, Glasgow.

THE BOOMERANG. From a broadside printed by Walker, of Durham. To be sung to the tune of *Weel dune, Cappy*. The song celebrates a fête at Dryburn Hall, the home of the Wharton family on the outskirts of Durham.

AW WISH PAY FRIDAY WAD COME. Written in 1870 by an Elswick miner named Anderson, to the tune of *Aw wish yor mother wad come*. Text from Allen's *Tyneside Songs* (Newcastle, 1891). A performance of the song by Louis Killen may be heard on a Topic record, 12T 189.

DURHAM GAOL. Words by Thomas Armstrong, directed to be sung to the tune of *Nae Guid Luck aboot the Hoose*. Mr W. H. Armstrong reported (1951): 'My father was wrongly sent to Durham Gaol for six months, and while in there he composed the song *Durham Gaol*. When the authorities saw it, they reduced his sentence.' Reduction of prison sentence following the composition of a ballad by the prisoner is a persistent piece of international folklore.

THE ROW BETWEEN THE CAGES. Words by Thomas Armstrong, directed to be sung to the tune of *Robin Tamson's Smiddy*. A performance of this ballad of a contest between machines may be heard on a Topic record 12T 122, sung by Louis Killen.

SCALLOPER JIM. This piece of epic hyperbole was written by an ex-miner, Joseph ('Jossy') Craggs of Durham (*c.* 1870–*c.* 1948), and published by him on a broadsheet, undated, in the early 1930s. A 'scalloper' is one who works coal entirely by hand, without the use of gunpowder. The strophic arrangement, such as it is, is by the compiler.

LITTLE BIT O' THIS. From Mrs Alma Wright, of Kippax, near Leeds (June 1951). 'Twist' is, of course, chewing tobacco, and Mrs Wright explained that 'whirlers' are 'old tops of stockings fastened round the boots to keep out the dust.'

MY GAFFER'S BAIT. A song made by John Pandrich ('Johnny Handle'), at a time when he was apprentice at the East Walbottle Colliery, Northumberland. Near sanctity attaches to a miner's 'bait' – the food he has brought with him to the mine – perhaps because it is a link with home. Hence the twofold embarrassment of the youngster who spoils the lunch of his boss, the mine surveyor. The song is sung by Johnny Handle on a Topic record, 12TS 271.

HALLY'S PIEBALD GALLOWA'. George Purdon made this one, and set it to the tune of *The Wearing of the Green*.

NE NETTIES DOON THE PIT. From the singing of Bert Draycott, of Fishburn, Stockton-on-Tees. Mr Draycott says: 'This one was written by Dennis Rowan of Elswick, and touched up a bit by myself.'

THE GRAVEYARD SHIFT. The song was made by Les Pearson, based on jokes heard from his workmates at the Rising Sun Colliery, Wallsend. He set it to the tune of *Rosin the Bow*. The term 'graveyard shift' was originally applied by American miners to the third (late night) shift. The term came to Co. Durham via Nova Scotia, where many north-eastern colliers worked in the past. In Scotland it is sometimes called the 'dying shift'.

AFTER A PIT PAY DAY. Communicated by John H. Denney, of Stainforth, Doncaster.

WAKE UP, WAKE UP, MY LOVE. A squib by Ned Booth of Birtley, Co. Durham, set to a popular Scottish melody for the ballad of *Willy O*, often called *The Night Visiting Song*.

PIT LIE IDLE. From the singing of Pat Elliott, at Birtley, Co. Durham, July 1963. Mrs Elliott says: 'It's been a family lullaby for years. All the ten grandchildren, to which we've contributed half, have been patted to sleep with this song.'

GLOSSARY OF MINING TERMS

Back-ower torn (back-over turn): A manoeuvre by which the putter would require to go past a turn, throw a switch, then back the pony and tubs around the turn, a hazardous and disliked operation (Elliott).

Back shift: The second shift of hewers in each day. It commences 4 hours after the pit begins to draw coals (Greenwell). In modern times, in some places coal may be loaded in back shift (late morning or afternoon); in others, during fore shift (early morning), and they may alternate. The shaft could not handle all the coal if the whole lot were loaded in the one shift (Elliott).

Baff Week: When wages were paid fortnightly, the Baff Week was the second week of the fortnight (miners were notoriously short of money during Baff Week).

Bait: A cold snack, usually sandwiches, taken underground by miners, called 'snap' in some areas.

Band: Slate or other rock interstratified with coal (Fay).

Bank: The surface.

Banksman: Man in attendance at the mouth of a shaft, who superintends the work of sorting and loading the coal (Gresley). Man who (a) draws the full tubs from the cages at the surface, when wound up by the winding engine, and replaces them with empty ones; (b) he also puts the full tubs to the weighing-machines and thence to the screens on which he teems the coals; (c) it is also his duty to keep an account of the quantity of coals and stones drawn each day (Greenwell). According to Elliott, (b) and (c) are part of the 'head-keeker's' job.

Bellman: Bailiff.

Belt: Conveyor.

Bleazer (blazer): A tin pan or sheet of iron, struck to attract attention, e.g. to give warning of a meeting. A sheet of steel about 3 ft × 2 ft, with handle attached, used for drawing up a fire. Struck with a poker it makes a loud noise suitable for attracting attention or for 'tin panning' (charivari). A married couple not on speaking terms are said to be 'chalking on the bleazer' (Elliott).

Bord (or '*Board*'): A wide bord is an excavation, a pillar (20–40 yds in length and 4–5 yds in width), usually driven at right angles to the cleavage of the coal; sometimes, however, when the coal is very flaky and works tender, it is advantageous . . . to drive the boards in the direction of the cleat, when they are called *headways boards* (Greenwell). A bord less than 4 yds in width is called a *narrow bord.*

Bordways (pron. 'bordrus'): The direction at right angles to the main cleavage planes (C. & M.M.P.).

Braffen: Horse-collar.

Bunker: Area of loose coal after firing shots. Sometimes called a 'jud'.

Byuts: Boots.

Ca': Drive.

Canch: Part of the roof which has to be taken down, or of the floor which has to be broken up, in order to equalize the grade of the road, or to permit the movement of tubs (Redmayne). Canch may also signify stone that has been removed to pack 'goaf', i.e. the part of a mine from which coal has been worked away, leaving a space that requires filling up (Elliott).

Candyman: Rag-and-bone man, scrap collector, from the practice of giving children candy in return for supplying saleable discards.

Chum 'uns (chummins, tyum 'uns): Empty tubs, from Sandinavian *tøm*, 'empty'.

Clarty: Dirty, muddy.

Cleat: The main set of joints along which coal breaks when mined. A stupid person is said to be 'brokken in the cleat' (Elliott).

Closer: Small piece of rail used to bridge gaps.

Corf (corve): Large basket used in hoisting coal; from the German *Korb*, a basket. According to Greenwell, the standard corf would be of hazel twigs, holding 10–30 pecks. Smaller corves, of 6–8 pecks, were used for stones or rubbish.

Crack: Conversation.

Cracket: A low wooden stool or seat upon which the hewer can sit in performing parts of his work (Greenwell). A three-legged cracket, with one leg shorter than the others, might be used for supporting the hewer's shoulder, for leverage while using the pick. It was then called a 'nicking styul', because it was particularly employed in nicking out the 4-inch undercut. A hewer could be recognized because of his 'nicking shoulder' – one shoulder higher than the other through contact with the 'nicking cracket' over the years (Elliott).

Deputy (depitty): Underground official who attends to the security of the roof and sides of the mine, over a certain number of stalls, and looks after the safety of the miners (Fay). Charge-hand in a district, responsible for a written report each day. A deputy might also fire shots. In fact, most deputies came from the ranks of the shot-firers (Elliott).

Doggy: Overman responsible for maintaining the rails on which the tubs run.

Double tokens: Attempts might be made to get paid twice for a filled tub by sneaking an extra token (q.v.) on the tub in the hope that it would not be noticed in the rush of drawing coal (Elliott).

Dreg (dregg): A piece or pieces of iron or wood put between the spokes of a tub or wagon wheel to check its progress where the dip of the way is considerable (Greenwell). Dregs might be round or conical. The word has passed into colloquial speech; 'Put a dreg in' means 'slow down' (Elliott).

Drift: Passage underground, following the vein of the coal. A heading driven on the strike of a coal seam (Fay). A long slow heading from one level to another (short sharp headings were called 'hitches') (Elliott).

Duds: Pit clothes.

Dues: A gratuity, sometimes money baked inside a small cake, as a Christmas gift reckoned to be due from filler or hewer to putter or driver (Elliot).

Engine plane: Underground way, either level or dipping inbye or outbye, along which the trams are conveyed to and from the workings by engine power (Gresley).

Flat (flatt): A district or set of workings in the mine. Sometimes a siding or station underground (Fay).

Forst (first): 'On forst', on first shift.

Fugleman: Leader, spokesman.

Gad: A steel wedge; also, a percussion drill.

Gallowa (galloway): A pit pony.

Gannin': A tunnel connecting the main road and the coal face. 'From gannin' doon te lowse', from starting time to knock-off time.

Gannin' board: Main tunnel.

Graveyard shift: The third, or night shift.

Grovet: Scarf (from 'cravat').

Headways (pron. 'headrus'): The direction of a place driven parallel with the way in which the coal breaks when mined; that is, head on (Gresley).

Hitch: A fault in the seam, giving a sudden change of gradient.

Hoggers: Stockings without feet, worn by miners. The word literally means 'hose'. As a pun, in the pits a water-hose is called a hogger. Latterly, hoggers were old long underpants or 'linings' (Elliott).

Holed (as 'thee place is holed'). Holed through to another working, necessitating a move to a new work-place.

Howerman, see *Overman:* The pronunciation is peculiar to West Durham and Cumbria.

Inbye: Towards the working face or interior of the mine; away from the shaft or entrance.

Jenkin (or junking): An opening cut into, or a narrow slice taken off a pillar in the room-and-pillar system of working coal (Fay).

Jig: A self-acting incline; a gravity plane underground (Fay).

Jowl: To knock on the roof with a pick-handle or hammer, in order to test its firmness. A sort of tattoo beaten with a hammer, and in response, upon the faces of two places or drifts near holing, or intended to hole into each other, by a person in each place, for the purpose of ascertaining by the sound their relative positions (Greenwell). Also used to contact men trapped in falls or cave-ins. The word has passed into general usage; to threaten to jowl someone implies a punch-up (Elliott).

Jud, Judd: A portion of the seam, kirved, nicked, and ready for blasting (Greenwell). A working place, usually 6 to 8 yards wide, driven in a pillar of coal (C. & M.M.P.). The term is also applied to an area of loose coal after shots have been fired. Or it may mean a block of coal about 4 yards square, holed and cut ready for breaking down (Fay).

Keeker: Man in charge at the bank top (Elliott). An inspector (probably from N. German 'kieken' – to look). Greenwell, however, gives the word as 'keeper', and defines it as an inspector attending the cleaning and screening of coal.

Keel: Flat-bottomed barge formerly used on the Tyne to carry coal to ships waiting in deeper water at the river mouth. Such barges were manned by *Keelmen.*

Kelly sweep (or *swape*): An operation similar to a 'back-over turn' (q.v.) but even more difficult to manage, because accomplished with moveable plates (Elliott).

Kelter: Money.

Kerve (or *kirve*)] To undercut (C. & M.M.P.). In the North-East, pronounced 'korve'.

Kevil, Cavil, Kyevil: Originally a lot drawn at stated periods, to determine the place in which the miner will work during the following period. Nowadays often applied to the work-place itself. According to Elliott, a putter would also draw his pony or 'gallowa' for the quarter-year by kevil. The method was to put chalk marks on one side of a saw or plank, and the name of a district or pony on the other side, invisible to the pitman. The lottery would be conducted by the deputy or overman. The man made his choice by rubbing a chalk mark, hence the colloquial expressions 'rubbing his cavil' and 'the worst one on the saw' (Elliott).

Kist: Wooden box or chest in which the deputy or the timberman keeps his tools. The kist is placed at the flat or lamp station and is something of a gathering point.

Layout: To set aside and discount trams of coal deemed to be improperly filled, or to contain too much stone, etc.

Limmers: Shafts by which the ponies draw the coal-tubs (Fay).

Lowe: Light.

Lowse: Loose. From a 'lowse place' the coal comes away easily. The term also implies the end of the working day ('fre gannin te lowse'=from start to finish). To the sense of release there is an ironic edge, as if turned loose like a beast of burden. Alexander Barrass described death and after as 'eternal lowse'.

Marra: Workmate.

Mence: Pride, also decency, good manners. Nowadays also used to denote commonsense and a feeling for what is proper, as 'That bugger's got nee mence' (Elliott).

Mothergate: Main passage in a district of mine workings (Fay).

Neuk: Corner of a working place at the junction of face with one side.

Odd 'un: It was common to link up an odd number of tubs, usually 13. There would be an even number of putters (loaders). So, as each putter was paid according to the number of tubs he filled, the one who went fastest and so gained the extra tub, could increase his wages.

Outbye: Toward the mine entrance, nearer to the shaft and hence further from the working face.

Overman: The foreman of the underground workings. Mining official next in rank below the manager or viewer, if there is no under-manager (Greenwell). In most mines of any size he would be in charge of a seam, and would have several deputies, face-workers, etc. under him (Elliott).

Ower tree, over tree: Engine beam.

Pack: Wall of stone, replacing filled coal for part of face. Also, according to Elliott, stone for maintaining roadways, packed along the face.

Pass-by: Section of double-track, with points, where the tubs may pass. Called 'landing' in Durham.

Pillar: A solid block or area of coal left in for support.

Plate: Small stretch of temporary rail.

Put, Putter: To 'put' is to convey coal – with or without a pony – from the working face to an assembly point further outbye. The work, notorious for frequent derailments, is usually done by young men called 'putters'.

Ramble: A shale bed on top of a coal seam, which falls as the coal is removed (Fay).

Rap her to bank: Signal by knocking to indicate to the winder that the men wish to be drawn up to the landing at the top of the shaft (the 'bank'). According to Elliott, for many years now, such signals have been given by electric bell, but the term remains. 'Rap hold' is still in use in connection with conveyor belts, etc., signifying 'stop'; though here again an electric system is used.

Ringer: Crowbar.

Ripping: The débris consequent on the roof or top being cut or blasted down (Gresley).

Riving: Hacking at the coal-face, without method.

Roll: Drum of a winding engine. Also, according to Greenwell, a species of hitch. The roof of the seam coming down into the coal without any corresponding depression of the thill, thus causing a rip.

Saa: Saw used mainly for cutting pit-props to length; also employed as an aid to cavilling, work-places being chosen by means of chalk marks on a saw.

Sark: Shirt. A pit sark would be a body-shirt, a full cotton vest used by the older generation of miners, which had a vogue as a main shirt among youngsters during the 'tie and dye' era (Elliott).

Screen: A construction of grate-bars or perforated sheet-iron used to sort coal according to size.

Scufflins: Small coal shavings left in heaps by the cutting machines.

Set: A train of coal-tubs. More rarely, the term is also applied to a timber frame used for supporting the sides of a shaft or other excavation (Fay). Sometimes written 'Sett'.

Shaft: An excavation, narrow compared with its depth, made for finding or mining coal (or ore), for raising the stuff mined, for lowering or hoisting men and material, or for ventilating underground workings (Fay). Also, the wooden handle of a pick.

Shot-stick: Stick or pole for ramming powder into drill-holes before detonation.

Shull: Shovel.

Slush: To fill with mortar, cement and the like, as the joints of a wall. More generally, watery mud.

Slush-cap: A bowler hat without brim, worn in the days before safety-helmets as protection, particularly against dripping slush (Elliott).

Staith: Platform at riverside for loading coal on the keels.

Standing: A place where work cannot continue. A work-place awaiting the next operation (Elliott).

Swally: A depression in the underground roadway, allowing water to collect.

Tally: Token (q.v.).

Tewing (or '*chewing*'): Tearing at the coal face, without due control.

Thill: Floor of a coalmine (Fay).

Timmer: Timber. Any wooden prop, post, etc., used to support mine workings.

Timmerin: The timber structure used for supporting the faces of an excavation during the progress of construction.

Token: Small metal tag or disc bearing the hewer's number, to be attached to the tub to show the official up top who had filled it. The putters also had tokens, to hang on each tub they drew outbye.

Tram: A low bogey-truck used underground. In the old days it was employed for conveying corves (Greenwell). Nowadays it is mainly used for carrying timber, rails, etc. (Elliott).

Trepanner: Coal-cutting machine.

Tub: A truck used for conveying coal.

Tummelled off the way (of a tub): Left the rails. 'Most often and most irritating' (Elliott).

Viewer: Colliery manager.

Way: Underground rails.

Weighing: Pressure of the roof. Called 'weighting on' in Durham, where 'weighing' would simply imply having the tubs weighed at bank.

Sources quoted:

C. and M.M.P.	*Coal and Metal Miners' Pocket Book*, 9th ed. 1904.
Elliott	Pete and Len Elliott, originally of Birtley, Co. Durham. Private communication.
Fay	Albert H. Fay, *A Glossary of the Mining and Mineral Industry*, 1920, reprint 1947.
Greenwell	G. C. Greenwell, *A Glossary of terms used in the Coal Trade of Northumberland and Durham*, 3rd ed. 1888.
Gresley	William S. Gresley, *A Glossary of terms used in coal mining*, 1883.
Redmayne	R. A. S. Redmayne, *Colliery Working and Management*, 3rd ed. 1912.

INDEX OF TITLES

INDEX OF FIRST LINES